Dick, Kerr & Co. Limited
Engineers and Contractors

London, Kilmarnock and Preston
A History of the Company 1853 – 1919

JOHN SHORROCK
C.ENG., M.I.MECH.E.

BAE SYSTEMS
INSPIRED WORK

CONTENTS

FOREWORD AND INTRODUCTION

For more than 100 years that part of Strand Road in Preston, known as West Strand, and over a distance of about 600 yards, has been bounded on both sides by the walls of a massive engineering works known for the greater part of that time as the Dick, Kerr Works. In 1919 these works became part of the newly founded English Electric Company, the works on the east side of the road continuing, initially, with the established business of tramcar manufacture, and those on the opposite side of the road, the West works, with the manufacture of large electrical power generation equipment for power stations and electrical traction machinery for tramways and railways.

During the Great War, 1914 – 1918, the Dick, Kerr Works was recruited by the Ministry of Munitions and contracts were placed for the manufacture of munitions, and in 1917 an order was placed for the manufacture of major sub-assemblies and assembly of Felixstowe Flying Boats; this was the first aircraft related work undertaken by the company.

From 1919, and now part of English Electric, they built one Fairey N.4 Atalanta, one Ayr M.3 and seven Kingston P.5 flying boats, together with four Wren S.1 light trainer aircraft. The English Electric Wren an ultra-light aircraft competed in the Daily Mail Light Aircraft Competition of 1923.

Between 1926 and 1938, there was a break in aircraft related activities and the Strand Road works continued with the manufacture of tramcars, rolling stock for electric railways and electrical power generation equipment. With war in Europe looming once again, aircraft work was taken up once more and continued through the Second World War with the manufacture of components for Handley Page Hampden and Halifax Bombers which were then assembled and flown from a new facility and airfield at Samlesbury between Preston and Blackburn. In 1944 and 1945 English Electric received contracts to produce a total of 300 Vampire jet fighters of which only 6 aircraft had been completed by the end of hostilities.

After the war there was an appetite to continue in the aviation business. Vampire production continued and in July 1944, W.E.W (Teddy) Petter was recruited to lead a design team with responsibility for developing new aircraft designs. The result was the design for a new bomber which would become the English Electric Canberra and in January 1946 a contract was received for four prototypes. In 1947 English Electric moved their research and development activities to Warton and experimental test flying commenced with a contract to investigate flight at high altitude using a Gloster Meteor Mk. IV; then in 1949 work began on a new high speed aircraft, the P.1., which would become the Lightning fighter.

In 1960 there was pressure from government for consolidation within the aircraft industry and the British Aircraft Corporation was formed, comprising a merger of English Electric Aviation Ltd., Vickers-Armstrong (Aircraft), the Bristol Aeroplane Company and Hunting Aircraft.

In 1977 nationalisation of the aircraft industry resulted in the merger of the British Aircraft Corporation, Hawker Siddeley Aviation, Hawker Siddeley Dynamics and Scottish Aviation, then in accordance with the provisions of the British Aerospace Act 1980 the statutory corporation was changed to a public limited company (plc), British Aerospace plc. By 1985 the British Government had sold all of its shares.

Finally, at the end of 1999, British Aerospace merged, initially, with Marconi Electronic Systems to found BAE Systems.

Meanwhile the works on the west side of Strand Road had continued mainly with the business of manufacturing electrical power generation equipment, electric traction equipment and the manufacture of electric and diesel-electric railway rolling stock. It was sold to General Electric in 1960 and continued with the manufacture of railway rolling stock until the collapse of GEC in the late 1990's when the works was acquired by the French company Alstom.

The history of the English Electric Company from its formation in 1919 to the present day has been extensively researched and recorded in numerous books and other publications. Its predecessor companies, including Dick, Kerr & Co. Ltd. have also had considerable coverage in many books and periodicals, although, and particularly in the case of Dick, Kerr, these concentrated, mainly, on their products with relatively little attention being given to the business background of the companies or the people who founded, financed and managed them.

About 60 years ago a publication entitled 'The Story of The English Electric Manufacturing Company' was published by Ray Abbott. This was followed up by a series of articles in the Company newsletter, 'English Electric and its People', subsequently brought together in a single publication 'Early Histories of Some Companies of the English Electric Group', published by the Company. In 1951, English Electric published 'A History of the English Electric Company' basic data compiled by Garraway Ltd., of London; a chronology of events from 1824 up to the formation of English Electric in 1918; also in 1951 'Electricity in Transport', recording the achievements of the English Electric Group in the application of electricity to transport. The authors and compilers of these publications did not enjoy the easy access to the vast quantity of documentation and particularly on line sources that we now enjoy and consequently whilst those works provide us with good historical background to the company they contain very little information regarding the founders of the company and their personal relationships or how and why the various elements of the company came about.

John Kerr JP c1903.

As recently as the 1950s a large number, and probably most, of the older employees of the Strand Road Works in Preston, if asked about their place of employment would, in all probability, have answered Dick, Kerr's, but who was, what or were Dick, Kerr? Today the company name is popularly known, to most people locally, as the founder of a famous ladies football team and little else, perhaps it is time to take a closer look at the origins and evolution of the company within the context of 19th C industrial expansion and particularly the development and growth of public transport systems in the latter part of that century and the first decade of the 20th C.

Towards the end of the 19th century an urban transport renaissance was taking place. All over the UK and in many other countries street surfaces were being torn up and tracks laid for a new mode of transport, the tramcar. Having started in America some forty years earlier the tram was late to arrive in the UK, held back by the lack of enabling legislation, but once restrictions had been lifted there was an almost insatiable appetite displayed by local authorities who early on vied to be amongst the first to introduce the tram to their streets, first horse drawn, then steam and finally the electric tram.

Initially demand was met by American and German tramcar manufacturers with facilities for final assembly in the UK. British enterprise soon recognised that there were good profits to be made from tramcar manufacture. At first the challenge was met by established British companies with the appropriate skills but quickly followed by new companies. The most successful of the companies who came new to the business was Dick, Kerr & Co. of London and Kilmarnock, already in business as manufacturers of railway locomotives and railway equipment; they had also become established, and with a sound reputation, in the laying of tramways. Although they came to the business of tramcar manufacture quite late, why, when the downturn came, and apparently against all the odds, did they survive whilst their competitors failed?

This account of the history of the Dick, Kerr Company, up to the formation of English Electric in 1919, makes use of all the earlier, and more recent, publications with further valuable information being retrieved from contemporary sources including newspapers and journals, correspondence and minute books housed in the various Record Offices and Libraries. Extensive use has also been made of electronic document and data recovery media to both corroborate the earlier research and to add further detail wherever possible. In addition biographical detail is presented with regard to the founders of the company, and their relationship(s), along with other senior contributors to its success.

There is no intention to comprehensively record every detail of the manufacturing or other business activities of the various component companies from which the final incarnation of Dick, Kerr and Company Limited finally emerged; it has, however, been necessary to give sufficient detail regarding their range of activities in order to clearly illustrate the significant milestones in the development of the company and their products both nationally and internationally. The structure of the company and how it was formed is covered in some detail in an attempt to build a clear picture of how the company developed and an understanding of those Victorian men of business who had the vision, enterprise and energy to lay the foundations of what would eventually become one of the most successful military aircraft producers in Europe and one of the largest employers, in the manufacturing sector, in the Preston area.

Dick and Kerr Family Tree

WILLIAM DICK
(Merchant)
B - C1787
D - 14th May 1832 (age 45), Maybole, Ayrshire.
M - 24th Dec. 1826, Penninghame

MARY BRUCE
B - C1806
D - 28th Aug. 1852 (age 46), Maybole Ayrshire.

JOHN KERR
(Manufacturer of Silk, Wool & Cotton)
B - C1821 Glasgow
M - 2nd July 1850 Barony, Lanarkshire

MARGARET (BRUCE) DICK
B - 8th Nov. 1828

WILLIAM BRUCE DICK
(Accountant)
Founder of W.B. Dick & Co
B - 23rd May 1830 Maybole, Ayrshire
D - 12th July 1905, Sevenoaks, Kent.
M - 2nd Oct. 1861, Elizabeth H. Harker

JAMES KERR
(Iron Merchant)
Kerr, Stuart & Co.
B - C1851 Glasgow
D - 4th Dec. 1884 (age 33)

JOHN KERR, JP, E. Lothian.
(Engineer)
DICK, KERR & Co.
B - 22Aug. 1852 Barony, Lanarkshire
D - March 1925 aged 72, Frinton-on-Sea, Essex.
M - C1877 Helen Margaret Melville
MP FOR PRESTON 1903 - 1906

A FAMILY BUSINESS

W. B. Dick

The Company known as Dick, Kerr and Co. and later as Dick, Kerr and Co. Ltd has its earliest roots in W. B. Dick and Co., founded in Glasgow in 1853, and its founder William Bruce Dick.

His sister, Margaret Bruce Dick had married John Kerr (Senior) in 1850. John Kerr was a partner in the Glasgow firm of McMillan and Kerr, Textile Manufacturers, with premises in Queen Street, and employing about 600 people, the partnership was dissolved on 1st December 1851. He then founded the firm of John Kerr and Co. (later More & Kerr), rivet and bolt manufacturers with premises also on Queen Street. Following the death of John Kerr, in 1856, William Bruce Dick takes over his partnership in the rivet and bolt business which then became More and Dick, then from 1859, after this partnership was dissolved, the business became a wholly owned part of W. B. Dick and Co. His sister Margaret Kerr retained an interest in the business.

The father of William Bruce Dick, William Dick a Merchant, died in 1832 when he was only two years of age; his mother did not remarry and died in 1852 at the age of 46.

William Bruce Dick was educated at the Ayr Academy and then articled to a firm of accountants in Glasgow and duly served his apprenticeship in the profession of accountancy. After completing his articles he became book-keeper for a big commercial house and subsequently joined a Mr. Reid as commission agents, but that partnership did not last long and was dissolved. He then commenced business in Glasgow, on his own account, as an oil refiner under the title of W. B. Dick and Co. in which enterprise he would become extremely successful.

Early documented evidence of William Bruce Dick as the proprietor of a business is in 1853, at the age of 23, when he founds W. B. Dick and Co. at Mavisbank Quay, Glasgow on a site subsequently to be used in 1888 for the Glasgow Garden Festival. Then in April 1859 W. B. Dick & Co., Oil Refiners, advertise -

> Store to let at 28 St. Enoch Square, with cart entry from Adam's Court Lane, a very commodious little store of ground flat and vaults, the latter being fire-proof. The store is well lighted, and possesses an excellent new crane capable of lifting two tons, and will be let at reduced rent by the present occupiers W. B. Dick & Co., Oil Refiners, 5 Dixon Street, in consequence of their removing to more extensive premises.

This move would appear to coincide with him taking sole control of the Rivet and bolt business of More and Dick.

In September 1859 he places an advertisement, in the Glasgow Herald advertising for a second hand Boiler –

> Wanted a good second-hand boiler 12 x 3½ ft. or thereby, apply W. B. Dick, Clyde Oil Works on Paisley Road.

In 1862, together with his sister, William Bruce Dick went into partnership with William Jackson in an oil refining business in Liverpool, which traded as William Jackson and Co. The partnership was terminated in October 1868 after which the business continued to function as part of W. B. Dick and Co.

The Glasgow Post Office Directory for 1864 – 65 provides the following further information on the diverse business interests of W. B. Dick and Co.

- Dick, William B. (of W. B. Dick & Co.), house, 21 Sardinia terrace, Great Western road.
- Dick, W. B., & Co., agents for David Corsar & Sons, sailcloth manufacturers, Arbroath; office, 17 Dixon Street.
- Dick, W. B., & Co., manufacturers of rivets, bolts, and spikes, Clyde Rivet Works, Mavisbank quay; office, 38 West Howard street.
- Dick, W. B., & Co., oil merchants and refiners, Clyde oil works, Spring Quay; 38 West Howard Street.
- Dick, W. B., & Co., fire engine manufacturers, and patentees of Dick's Portable Fire Engine L'Extincteur, 38 West Howard Street.

This clearly indicates that, even at this early period in the development of the company, William Bruce Dick is employing a policy of diversification as a business strategy and would appear to be constantly seeking new business opportunities.

By 1864 W. B. Dick is clearly adding to his range of manufactured products when he starts producing a man portable fire extinguisher, the invention of Monsieur Vignon, an Engineer Officer in the French Army. In January 1866 there is a report of 'the first experiment in Dundee' by Mr Dick, assisted by Mr Stark, the local agent; the ground selected was the site for the Albert Institute -

> A huge pile of wood and barrels upwards of six feet high was erected on the ground. To anoint this mass of material seven gallons of tar and a large quantity of spirits of petroleum were used, and the whole, when kindled, made a fearful blaze. When the pile was in an entire state of combustion, Messrs Dick and Stark advanced, each carrying a small cylinder on his back. Attached to the cylinder was a small flexible tube with a brass nozzle, the orifice of which was scarcely a quarter of an inch in diameter. Each gentleman, by merely turning a small tap, sent what seemed only a tiny stream of water all over the burning mass, and in less than three minutes reduced it to a black smoking heap of charred sticks.

> The next experiment involved a tank 18 feet long by 3 feet broad and 3 inches deep filled with tar, over which was poured five gallons of spirits of petroleum. This compound was set fire to, and, when thoroughly lighted produced so fierce a blaze that it was impossible to stand within several yards of it. Nevertheless, the streams from the cylinders being directed on the tank, the flames were extinguished in thirty seconds.
>
> Both experiments gave the highest satisfaction to the spectators, and many gentlemen who were present warmly congratulated Mr Dick on the merits of the invention. The Duke of Wellington was so convinced of the value of the invention that he supplied all the farmers on his estates with the cylinders. This was imitated by the Duke of Sutherland and the Prince of Wales fixed an early day for having experiments made in his presence.

In February 1866 there was an advertisement in the Northern Whig under the heading 'Safety from Fire'-

> The new Patent Portable Self Acting Fire Engine L'Extincteur. The above machine affords a hitherto unknown protection from fire. It is warranted harmless to life, health, or property; is as easily handled as a bucket of water, and will extinguish any fire in its early stage. For Factories, Warehouses, Ship Yards, and Stores it is of the utmost value; and on board Ships, in Country Mansions, Farm Houses, and Stock Yards, where no extraneous assistance is procurable, it is invaluable. Prices ranged from £2 12s 6d to £6 0s 0d; and, in copper, and for gentlemen's mansions, from four to eight Guineas. Messrs. W. B. Dick & Co., 36 & 38, Howard Street, Glasgow, sole Makers and Representatives in Scotland and Ireland for the Licensee.

The extinguisher clearly meets with the approval of the customers with the publication of a number of Testimonials, quote – to Mr. James Sinclair, April 16, 1868.

> Sir, having to-day had an excellent opportunity of seeing the operation of L'Extincteur, I have very much pleasure in saying that it more than fulfilled any expectation I had been able to form of its efficacy. It has a magical effect on the fiercest flame, and appears simple enough to be used by any person even of a very moderate intelligence. – I am yours, very truly, James Law, Mayor of Bradford.

Reported on the 5th April 1873 –

> An 'experiment' with Dick's chemical fire engine L'Extincteur, took place on the beach at Hastings, a few yards to the west of the Pier. This had been advertised in the local press and as a result, it was estimated, attracted from 1,000 to 2,000 spectators, including the Mayor. The demonstration, which was performed by Mr. Sinclair and an assistant involved setting fire to a small wooden building 16 ft. long by 4 ft. wide and about 8 ft. high; the flames were 'totally subdued' in about 30 seconds.

Advertisement from 1866

> In the seven years since it was introduced 'no less than 3,000 of the engines had been sold in this country and America, and the demand is increasing'. A short time before, Mr. Sinclair had visited Egypt 'where he was most favourably received by the Khedive and had received an immediate order for 400 L'Extincteur'.

The fire extinguisher business continues to expand; new models are added to the range with manufacturing and marketing being handed to a specialist company, the Fire Appliance Company.

In April 1890 a demonstration is staged on the beach at Douglas, Isle of Man before a large crowd including the captain of the Douglas Fire Brigade.

> The tests were under the superintendence of Mr. John Beesty, the Manchester representative of the London Fire Appliance Company (the proprietors and manufacturers of Dick's Extincteurs), assisted by the local representative, Mr. James Cubbon. --- The extincteurs tested were the "Fire Queens" and the "Lion" tubes, or hand fire extinguishers. --- They vary in size and in price, from the "Express Chemical Fire Engine" mounted on wheels, to the "Little Fire Queen," and the "Lion Extinguishers," which are hand tubes. --- They are kept ready for use at Windsor and other English Royal residences, and they have long been in use at the stations on the Lancashire and Yorkshire railway. At the various exhibitions the extincteurs have gained successively ninety six first prize exhibition honours, but the best proof that their merit as prompt fire extinguishers is becoming generally known, is that already 158,000 of them have been sold.

A brochure produced by The Fire Appliance Co. (Dick's Patents) in 1911 quotes testimonials from the Manchester Fire Brigade and such worthies as the Duke of Argyll, the Duke of Beaufort, the Marquis of Waterford and Lord Charles Beresford. The range of Dick's Fire Extinguishers listed and illustrated in the brochure includes –

> Dick's "Fire Queen" size 40 gallons of water "Double Express" mounted on wheels, price £48-10-0.
>
> Dick's "Fire Queen" sizes 10 to 25 gallons of water "Single Express" mounted on wheels, No's 1 to 3 prices from £14-5-0 to £27-10-0.
>
> Dick's "Fire Queen" sizes from 5½ to 9 gallons, prices from £5-0-0 to £7-0-0, over 300,000 sold.
>
> Dick's "Domestic Fire Queen" sizes from 1½ to 4 gallons, prices from 75/- each to 100/- each.
>
> Dick's "Little Fire Queen" sizes from ¾ to 3 gallons, prices from 16/- each to 28/- each.
>
> The "Lion" Hand Fire Extinguisher priced at 48/- per dozen.
>
> The "Lion" Hand Grenade priced at 36/- per dozen.

(16)

Dick's Fire Extinguishers.

DICK'S "FIRE QUEEN."

"DOUBLE EXPRESS."

THIS is an enlarged and more powerful engine of the "Single Express" type, and consists of two Cylinders, each measuring $42\frac{1}{2}$ inches high by $14\frac{1}{2}$ inches in diameter, which are fitted upon Carriage with 30-inch Wheels. It holds about 40 gallons of Water, and generates and discharges in action about 400 gallons of Carbonic Acid Gas. It has Charge Box for two extra charges, two 15-feet lengths of $\frac{3}{4}$-inch best Hose, coupled in centre, and a 10-inch Brass Director.

This is the largest and most efficient Chemical Fire Engine in existence, and for Colleges, Schools, Hospitals, Churches, and all large buildings where it can be handled is the surest protection against serious fire possible.

Price complete **£48 10s.** Charges **12/6** each.

STRONGLY RECOMMENDED TO COLONIAL SHIPPERS.

Dick's Fire Queen, "Double Express".

Dick's Patent Fire Extinguishers. Awarded 104 First Prize Exhibition Honours. Adopted Extensively All Over The World.

In the early days William Bruce Dick appears to have personally examined the company accounts, a task for which he was professionally trained and well qualified. Reported, in Liverpool, on 1st February 1864 –

> A young man of respectable appearance was charged with having embezzled several sums of money belonging to his employers, Messrs. W. B. Dick and Co., oil refiners of Glasgow. The prosecutors have a branch office in Oil Street, in this town, of which the prisoner had the management. On Monday last Mr. Dick came to Liverpool, and examined the books of which the prisoner had charge, he found several discrepancies which caused him to make enquiries. He then ascertained that on the 8th January Mr. Hollingshead, cashier to Messrs. T. Royden and Sons, paid to the prisoner on account of his employers the sum of £619. The prisoner, however, had only accounted for £615. There being several other cases against the prisoner, he was remanded until Friday next.

Again in Liverpool and reported on 24th January 1865 –

> A well dressed young fellow, apparently about 18 or 19 years of age, was charged with having embezzled two sums of money, amounting to between £11 and £12, belonging to Messrs. Dick and Co., oil refiners, &c., Mavis Bank, Glasgow. The prisoner, who was employed as a clerk in the Liverpool office on Oil Street, Liverpool, was remanded for seven days, so that the prosecutors' books might be examined. At a later hearing, reported on 28th January, he was committed for trial at the sessions.

In 1869 W. B. Dick & Co. moved their operations to a new purpose built works in York Street, Glasgow. The Glasgow Trades Directory for 1869 has an Entry for Dick, W. B. & Co. 15 York St. Makers of the new patent self-acting fire engine, L'Extincteur. But within two years disaster strikes when on 8th February 1871 the premises on York Street are destroyed by fire –

> A fire of a most destructive character occurred in the premises in York Street occupied by Messrs. W. B. Dick & Co. an oil refinery. It was evident from the first that the tenement in which the fire originated could not be saved, and the efforts of the firemen and the large posse of policemen present were mainly directed to preventing the further extension of the flames. So rapidly did the fire spread that the contents of the refinery were soon completely destroyed, and large portions of the walls gave way. The buildings destroyed consisted of two large blocks situated at No's, 9 and 15 York Street. The whole building, consisting of four storeys, had a frontage of 100 feet to York Street, with a breadth of 60 feet. Behind was an open court about 20 feet wide, and beyond was the oil refinery, erected about two years previously by the proprietors. The building had a 'vaulted sunk flat' in which were tanks for containing the oil. In the refinery there was a large quantity of oil and machinery, the tanks in the sunk flat were filled with the refined material. There was also in the premises a number of L'Extincteur fire-engines, for which Messrs. Dick were agents and patentees; and a lot of canvas, the manufacture of Messrs. D. Corsar & Co., of Arbroath, for which W. B. Dick also acted as agents. All the books and other valuable documents belonging to Messrs. Dick were preserved in one of Milner's safes, having fallen from the top storey; the entire stock of oil was destroyed.

For insurance purposes the machinery and material in the refinery were estimated to be worth about £10,000, and these together with the value of the buildings, and damage sustained by other parties, brought the loss to about £80,000.

Fire damage seems to have been a recurring problem to the company, although, considering the nature of the business, perhaps more of an Industrial risk. Reported on 3rd July 1893 –

> Yesterday afternoon a fire broke out at the works of Messrs. W. B. Dick & Co., oil refiners, Crownpoint Road, Glasgow, the damage done amounting to about £400. Again, on 16th October 1899 under the headline 'Big Fire in Glasgow' oil refiners' works in Flames. On Saturday forenoon an outbreak of fire occurred in the premises of Messrs. William Bruce Dick & Co., oil refiners, Glasgow. --- The part of the building in which the fire originated was a three-storey building filled with tanks containing oil in process of refinement. The roof collapsed although the oil contained in the tanks was not so far advanced in refinement as to be inflammable, but large quantities of it were destroyed. The damage was 'expected to be very heavy'.

James and John Kerr

In 1875, the nephews of William Bruce Dick, James and John Kerr, are in business as Iron Merchants and suppliers of Railway and Colliery plant. They are sharing premises with William Bruce Dick, oil merchant, at 22 Hope Street, Glasgow, thereby laying the foundation for a business partnership that would profoundly influence their future career paths and business strategies.

John Kerr became, in 1878, the Manager of the London Office of what was later to be Hurst, Nelson & Co., rolling stock builders. This venture seems to have been little more than a temporary diversion for by 1881 he is in partnership with his brother James.

In 1881 James Kerr founds James Kerr & Co. for the purpose of building railway wagons. The company seems to have quickly run into problems for in the same year the partnership Kerr Bros., Nelson & Co., is founded, when Andrew Sclanders Nelson becomes a partner, then Kerr Bros & Co. in partnership with John.

Reported in the London Gazette on 19th April and 27th May 1881 we find under The Bankruptcy Act, 1869 in the London Bankruptcy Court, -

> In the matter of Proceedings for liquidation by arrangement or composition with Creditors, instituted by James Kerr and John Kerr, of 175, West George-Street, Glasgow, and 101, Leadenhall-street, in the city of London, Iron Merchants, formerly trading as James Kerr and Co., at St. Enoch-square, Glasgow aforesaid then in partnership with Andrew Sclanders Nelson, as Kerr, Bros., Nelson & Co., at 175, West George-Street aforesaid, and 110, Cannon-street, in the said city of London, and lately as Kerr Bros. and Co., at 175, West George-Street aforesaid, and 101, Leadenhall-street, aforesaid.
>
> Notice is hereby given, that a First General Meeting of the creditors of the above named persons has been summoned to be held at the Cannon-street Hotel, Cannon-street, in the city of London, on the 6th day of May, 1881, at two o'clock in the afternoon precisely. – dated this 11th day of April 1881.

Further to the above -

> Notice is hereby given, that a First General Meeting of the separate creditors of the above-named James Kerr has been summoned to be held at the Cannon-street Hotel, Cannon-street, in the city of London, on the 6th day of May, 1881, at four o'clock in the afternoon precisely. – dated this 11th day of April 1881.

The experiences gained and lessons learned by these adventures into the business world would no doubt prove valuable in the future and appear to have been only a temporary setback, for in 1881 William Bruce Dick decides to separate the oil refining and rolling stock parts of his business by creating the Rolling Stock and Railway Materials Department of W. B. Dick & Co., managed by John Kerr. This decision was taken during the period that the company was engaged on the construction of the Alford & Sutton and the Ipswich Tramways; their earliest recorded joint projects, and an opportunity to test the effectiveness of the partnership. William Bruce Dick was prepared to invest his own capital in these ventures, demonstrating confidence in the engineering talent of the young John Kerr whilst at the same time recognising a good business opportunity.

The Alford & Sutton Steam Tramway

In 1880 a Bill was put before Parliament for the construction of a tramway of 2 ft. 6 in. gauge to run between Alford and Sutton-le-Marsh serving the villages of Bilsby, Markby and Hannah en route.

Throughout the route the public roads were to be followed, and at all save one of the passing loops there was to be a width of less than 9 ft. 6 in. from the edge of the footpath to the nearest rail of the tramway for a distance of 30 ft. or more. One of the clauses in the Bill was one to permit the tramway to convey goods, mineral and other traffic as well as passengers.

A delay of more than two years was to elapse before construction of the tramway could be started, and consequently it became necessary to apply to parliament for an extension of time. The Alford & Sutton Tramway Act was duly passed on 10th August 1882 and work was started on 14th December 1882.

The construction of the tramway was undertaken by W. B. Dick & Company and Mr. William Bruce Dick, was also the owner of the tramway. The design of track adopted was known as 'Kerr's second system' and designed by John Kerr; this consisted of grooved girder rails, 4½ inches high and weighing 45 lbs per yard. These were laid to the authorised gauge of 2 ft. 6 in. on transverse concrete sleepers 3 ft. apart between centres. Steel tie plates, ¾ inch thick, were laid on and embedded in the sleepers, the road being macadam. There was no paving, except at passing places, where 4 inch cubes were laid.

After a number of construction problems, identified by a Board of Trade inspection in July and December 1883, and the need to address complaints from local residents, trades-people and other vested interests, the tramway was eventually completed. In March 1884 the tramway was again inspected by Major-General Hutchinson on behalf of the Board of Trade and after journeying over the whole of the line he announced that he was perfectly satisfied and he officially passed the undertaking. In the evening Mr. Dick, the constructer and owner of the line, entertained to dinner the whole of the highway surveyors of the different parishes through which the line ran, a few friends and the representatives of the Press.

In his final report Major-General Hutchinson criticised locomotive No. 1, which had been in use since July 1883 hauling construction materials, as not being fit for passenger traffic.

He recommended that two additional brake blocks, with double couplings be provided with some method by which the driver could operate the car brakes. On the day he wrote his report – 8th March 1884 – he had inspected the almost completed locomotive No. 2 at Merryweather's Greenwich works. This, he said, was a very well arranged engine, "one of the best adapted I have seen for working from either end".

The tramway was to be formally opened for use by the public on 2nd April 1884. On the day before the opening a successful trial run was made by locomotive No. 2 driven by J. Compton Merryweather, assisted by C. J. W. Jakeman, CE, of London. The locomotive emitted hardly any smoke, condensed steam or noise and it was observed that passing horses were not perturbed. The engine drew a total load of 25 tons, 'equal to four London tramcars'.

In the evening a public tea was provided in the Corn Exchange followed by a public meeting at 9 p.m. at the end of which the chairman proposed success to the Alford & Sutton Tramway, a toast which was responded to by Mr. Dick, who said "he believed that the undertaking would prosper, although they might have to wait a year or two"; the cost of the tramway had exceeded £30,000. A further bonus and lasting legacy from the building of the tramway was that the roads were in a better state than they had ever been. In 1885 the tramway purchased its third and last engine, No. 3, including rolling stock, from Dick, Kerr and Co. at Kilmarnock.

With the expansion of the Great Northern Railway in 1888 the Alford & Sutton Tramway was unable to compete and became unprofitable and effectively ceased operations in December 1889.

South Staffordshire and Birmingham District Steam Tramway

In the same year W. B. Dick and Co. of Leadenhall Street, London were contractors for the South Staffordshire and Birmingham District Steam Tramways; 'representatives at the launch ceremony were W. B. Dick and Mr Kerr'. The Staffordshire Tramways Order of 1879 authorised the construction of the steam tramway which was operated by the South Staffordshire and Birmingham District Steam Tramways Company. The company depot was at Kings Hill, Wednesbury. On 26th December 1882 a line was opened, on the 3 ft 6 in gauge, between Aston Manor and Old Square, Birmingham by the Birmingham and Aston Tramways Company Ltd. This was followed, between 1883 and 1885, by a network of narrow gauge lines in the Black Country and operated by the South Staffordshire and Birmingham District Steam Tramways Company Ltd, which met the existing horse lines at Handsworth.

The Engineering and Metal Trades Exhibition of July 1883, and reported in the Engineer of 6th July, had an exhibit from Messrs. W. B. Dick and Co., of Leadenhall-street, of a locomotive for the 2 ft. 6 in. gauge, intended for Burma. The most noteworthy feature, of the locomotive, was the valve gear which 'closely resembles Joy's, a swinging link taking the place of the curved incline'; the report included an illustration of the valve gear. Also –

> Various portions of Kerr's patent portable railway are exhibited by Messrs. W. B. Dick and Co., of Leadenhall-street, E.C., together with specimens of rolling stock and fittings. Like other portable railways, it has been designed for use on farms, plantations, iron mines, &c., and in all cases where simplicity and quick mounting and dismounting of the line are desired. The sleepers are all made perfect to gauge, securing accuracy in laying without having to resort to skilled labour, and the gauge can never vary unless the rails are absolutely torn away from the sleepers. The rolling stock comprises covered and open goods wagons, equilibrium and universal tipping wagons, and carriages for timber and sugar cane, as well as passenger cars; while among the fittings will be found points and crossings, switches, wheels and axles, and various types of fixed and portable turntables.

The General Directory for Glasgow for 1885/86 provides further information on the activities of the company at that time –

- Dick, W. B. & Co., Limited, agents for David Corsar & Son's sail cloth manufacturers, Arbroath; office, India court, 22, Hope Street.
- Dick, W. B. & Co., Limited, Paint manufacturers and sole proprietors of Hannay's patent paints for coating ships bottoms, Clyde Oil Works, Crownpoint road, office, India court, 22, Hope Street.
- Dick, W. B. & Co., Limited, oil merchants and refiners, paint manufacturers, Clyde Oil Works, Crownpoint road, office, India court, 22, Hope Street; branch establishments at London, Liverpool, Manchester, Barrow-in-Furness, Newcastle-on-Tyne, and Cardiff.
- Dick, W. B., fire engine builder and patentee of Dick's portable fire engine L'Extincteur; works, Crownpoint road, office, India court, 22, Hope Street.

The various Dick products appear to be giving satisfaction to their customers, who seem happy to publicise their experiences, no doubt to the mutual satisfaction of both parties with the added bonus of free publicity for Messrs. W. B. Dick and Co; the following are a few typical examples –

> The fastest passage on record from Cape Town was a headline on 5th February 1889. The S.S. Roslin Castle, of the Castle Mail Packet Company; actual steaming time, calling at St. Helena, Ascension, and Madeira was 17 days, 6 hours, 40 minutes. The ship was coated with Hannay's Composition, manufactured by the proprietors, W. B. Dick and Co., Limited, 20, Bucklersbury, London, E.C.

On Friday 30th October 1891 the weekly meeting of the Mersey Docks and Harbour Board confirmed a recommendation to accept 'the tender of Messrs. W. B. Dick and Co. to supply 4000 gallons of mineral oil for use at the lighthouses'.

As early as 1876 W. B. Dick and Co. are found supplying olive oil to the Mersey Docks Board, on the recommendation of the marine committee.

On Wednesday 19th June 1895 there was a report on the arrival at Southampton of the liner Louis at the end of her maiden voyage from New York. –

> Captain Randle reported 'the engines ran sweetly throughout, and there was never a sign of heating in the bearings'. The Liverpool firm of W. B. Dick and Co., of the Mersey Oil Mills, supplied the Louis with their compound engine oil, and within the following few days were expected to despatch some 2000 gallons to Southampton for use on board.

Along with the good reports there were the occasional publicity setbacks. A Board of Trade inquiry reported on Wednesday 6th January 1897 into an explosion on board the steamship Scotia, of London, in the English Channel, near the Eddystone, on the 8th October 1896. –

> The Scotia was owned by the Telegraph Construction and Maintenance Company (Limited), and was used for the purpose of carrying and putting down submarine cables. On 6th October she sailed from the Thames bound for Vigo, and had on board were some drums of Messrs. W. B. Dick and Co's anti corrosive paint, or 'boat-topping', which was stored in the forepeak. On nearing the Eddystone, when one of the crew was going down the forepeak with a naked light, an explosion took place, with the result that the bows of the ship were seriously damaged, and she had to put into Plymouth for repairs. The question for the court was as to how the paint came to be stowed in the forepeak, which was not ventilated, and further as to how it escaped from the drums. The Board of Trade also desired to know whether the paint was inflammable, and if so, whether Messrs. Dick and Co. gave notice to that effect. Apparently the painting of the ship had not been completed when Captain Cato the master of the Scotia received orders to get the ship ready for an early departure and a drum and a half of paint that remained was stowed in the forepeak.
>
> Captain Vivian, manager of the Paint Department of Messrs. W. B. Dick and Co. said that "before he heard the evidence of Dr. Dupré he had no idea that this paint was of an explosive nature, though he knew that it was inflammable".
>
> Mr. Marsham, in the course of his judgement, said that "Messrs. Dick and Co.'s patent boat-topping when exposed to the air is liable to give off gas or vapour at ordinary temperatures, and when mingled with air it is explosive and very dangerous". No notice or caution was issued by Messrs. Dick and Co. with the drums in question. No blame was attached to Messrs. Dick and Co. or to the master

of the Scotia, but the Court, though it did not positively blame the chief officer (Joseph Edmund Leach), considered that he was somewhat neglectful of his duty in not giving instructions for the stowage of the paint.

This latter event occurred during the period that W. B. Dick and Co. was in the process of incorporation.

W. B. Dick & Co. Ltd

When W. B. Dick and Co. was incorporated in 1896, the report in the Times newspaper for 12th and 14th December 1896 provides the following background information for prospective shareholders. –

> The business of this Company was established in Glasgow in 1854 by Mr William Bruce Dick and was so successful that in 1862 branch works were established at Liverpool and in 1875 in London, with depots at Newcastle-on-Tyne, Cardiff, Barrow-in-Furness, and Hamburg. Mr W. B. Dick carried on the business personally until 1886, since which date it has been conducted as a private corporation, the whole of the shares practically being held by him, and the present Company has been incorporated to acquire the business on and from the 30th June 1896.
>
> The business is that of oil refiners and importers, and manufacturers of antifouling paint for coating ships bottoms, carried on at the King and Queens Wharf and Factory , and Lavender Wharf, Rotherhithe, London; the Clyde Oil Works, Crownpoint Road, Glasgow; the Mersey Oil Works, Cotton Street and Carlton Street, Liverpool; at Barrow-in-Furness, Newcastle-on-Tyne and Cardiff; and at Hamburg.
>
> The Company commences with a very large connection, extending over the whole of the United Kingdom and abroad, as may be gathered from the fact that the number of their customers exceeds four thousand. This connection lies principally with the railway companies, steamship owners, mill owners and engineering works, to whom they sell their engine and burning oils, which oils have for many years borne a high reputation for quality. They also have a large trade in fish oils amongst soap boilers and tanners. The anti-fouling paints or compositions manufactured by the Company have for long been largely used by ship owners and recently, after three years experimenting and testing, the Admiralty have given orders to the Company for the coating of a considerable number of Her Majesty's ships. This fact of itself is calculated to be of great value to the Company in the future.
>
> The Directors of the Company are named as -
>
> Wm. Bruce Dick, Esq., Merchant, London, Managing Director.
>
> Bruce Dick, Esq., Merchant, London.
>
> Charles E. Dick, Esq., Merchant, Liverpool.

William Hay Hannay, Esq., Merchant, Glasgow.

John Kerr, Esq., Learmonth Terrace, Edinburgh.

Average annual profit earned over a period of six years was given by the Accountants to be £15,775 13s 4d and the value of properties and investments recorded as £135,572 14s 9d.

William Bruce Dick agreed to act as General Managing Director for at least five years, and Mr. Bruce Dick, who with his father was manager of the London business for the previous twelve years, consented to act for the same period. Mr. W. H. Hannay, the Manager at Glasgow, Mr. Rogers, the Manager at Barrow, Mr. William Robinson, the Manager at Liverpool, and Mr. Ketting at Hamburg, all of whom had conducted the Company's business at their respective houses for the previous ten years agreed to continue their services. 'It will thus be seen that the hitherto successful management and organisation will be continued'.

William Hay Hannay was managing director of W. B. Dick and Co. Ltd., from the mid 1880s until 1896 when the company was incorporated, after which he remained a Board member. In 1879-80 he was connected with the Clyde Oil Works of W. B. Dick and Co., on Crownpoint Road, Glasgow then in 1889-90 he is managing director. In 1889 he was also connected with -

- Hannay's Patents Co. (Lim) makers of Hannay's patent 'Electrogen' for preventing corrosion and scale in steam boilers; makers of white lead, paints and colours; Hannay's patent 'Electroid' anti-corrosive and anti-fouling compositions for preventing corrosion and fouling of ships' bottoms; also compositions and enamels for gas works, iron and steel structures, corrugated iron, stone work, etc; office, 67 Great Clyde Street; works, Caledonia Works, Possilpark; T. W. Duncan, Secretary; also,
- Hannay's patent Electroid antifouling composition, sole makers, Hannay's Patents Company (Ltd.), 67 Great Clyde Street.
- Hannay's anti-fouling and anti-corrosive compositions for ship's bottoms, white zinc and tallow for sailing vessels, etc; Dick & Co., Limited, India Court, 22 Hope Street, Glasgow.

The Glasgow Post Office Directory for 1895 gives W. H. Hannay as managing director of W. B. Dick & Co., Ltd., office 24 Queen Street; then in 1909 a director of W. B. Dick & Co., Ltd.

William Hay Hannay, Ex. Bailie of Glasgow, died on 27th August 1920, aged 69; his obituary dated 28th August 1920 reports the death of ex-Bailie William Hay Hannay, Glasgow, at his residence in Montgomerie Crescent after a prolonged illness. –

> Mr Hannay was closely associated with the firm of Messrs. W. Dick & Co., oil merchants and refiners, whose service he entered over 50 years ago. He was the elder brother of the chemist James Ballantyne Hannay, a prolific inventor who published several scientific papers and took out over seventy patents in Britain, Europe and the USA; James Ballantyne Hannay also formed his own Patents Company in Glasgow.

James Ballantyne Hannay's most controversial scientific work, also his best known was related to his claim, made in 1880, that he had successfully synthesised diamonds. On 26th February 1880 a paper on his process for producing diamonds artificially was presented to an 'unusually large gathering' of the ordinary weekly meeting of the Royal Society, London. –

The hall in Burlington House was completely filled, Mr. Hannay was not present and the paper was read by the Secretary.

In 1885 Hannay's Patents Company introduced the 'Lucigen' illuminator a new method of lighting, 'which bids fair to become a powerful rival of the electric light in its adaptability for illuminating large open spaces'.

The Glasgow Post Office Directory for 1889 – 90 contains the following with reference to the business activities of James Ballantyne Hannay -

- Hannay, J. B., F.R.S.E., F.I.C., patentee; private laboratory, 67 Great Clyde Street; works, Caledonia Works, Possilpark; res. Cove Castle, Loch Long.
- Hannay & Pape, patentees and sole makers of the marine white, 67 Great Clyde St., and Caledonia Works, Possilpark.

The Hannay archive is held by the University of Dundee.

On Wednesday 10th April 1899 there was notice of legal proceeding being taken by W. B. Dick and Co. (Limited) to protect its commercial interests under the headline Hannay's Anti-Corrosive and Anti-Fouling Compositions for Ships' Bottoms –

> Messrs. W. B. Dick and Co. (Limited), of London, Liverpool, and Glasgow, the assignees of the letters patent No. 770, of 1880, granted to Mr. James Ballantyne Hannay, of Glasgow, for a new or improved Anti-fouling and Anti-corrosion compound or mixture, applicable for preventing the fouling of ships' bottoms and other submerged structures hereby give notice that they have commenced an action against the said James Ballantyne Hannay, for an injunction to restrain the said James Ballantyne Hannay from manufacturing, selling, or using Anti-fouling and Anti-corrosion compounds manufactured in the manner protected by the said letters patent, and W. B. Dick and Co. further give notice that they will take legal proceedings against all persons manufacturing, selling, or using Anti-fouling or Anti-corrosive compounds that infringe the said letters patent.

The geographical spread and business interests of W. B. Dick and company in 1905, the year that the founder, William Bruce Dick, died is summarised in the Scottish Post Office Directory for Glasgow:

- Dick, William B., (of W. B. Dick & Co., Lim.), residence Carrick Grange, Sevenoaks, Kent.
- Dick, W. B. & Co., Limited, paint manufacturers and sole proprietors of Dick's patent paints for coating ships' bottoms, Clyde Oil Works, Crownpoint Road; office, 24 Queen Street.
- Dick, W. B. & Co., Limited, oil merchants and refiners, paint manufacturers, sole proprietors of Dick's (late Hannay's) patent anti-corrosive and anti-fouling compositions for coating ships' bottoms, Clyde Oil Works, Crownpoint Road; office, 24 Queen Street, and at 33 Eastcheap, London, E.C.; Cotton Street, Liverpool; Hindpool Road, Barrow-in-Furness; 20 Broad Chare, Newcastle-on Tyne; 1 James Street Docks, Cardiff, and Kleiner Grasbrook, Rethwisch 20, Hamburg.

Also in 1905 the Snodland (Kent) Historical Society Newsletter, has an entry for W. B. Dick & Co., Oil Refiners, having taken over a factory in Church Fields, with frontage on the river Medway. This seems to have been a short lived enterprise since by 1915 the premises have been taken over by the British Lianosoff White Oil Co. Ltd., oil refiners.

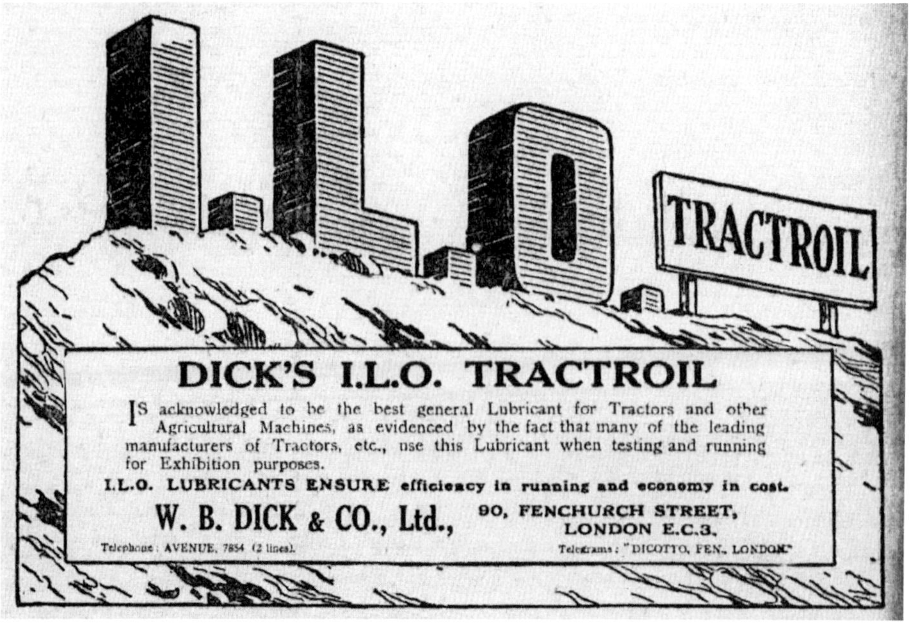

Advertisement from 1918.

In 1914 the business of the W. B. Dick & Co. is described as oil refiners, boilers, importers and rectifiers, distillers of turpentine. Specialities: marine engine and cylinder oils, turbine engine oil, crank chamber oil and fire-boiled oil for sheet dressing.

Quote from the British Commerce and Industry 1934 Review –

The history of W. B. Dick and Co. Ltd. had coincided with the scientific study of oils from mineral sources and the provision through analytical research of such oils to meet the new needs of the engineering world. Mineral oils were almost unknown and of academic interest only in 1853. The first Dick products, therefore, were all of vegetable or animal origin. The record of the company in effecting the change to mineral sources placed it in a leading position among world refiners and blenders of mineral oils, and the careers of the chairman and managing director, Mr. W. Bruce Dick, and of Mr. C. E. Dick, also managing director, both sons of the founder, give them the longest individual connection with this new industry in the British Isles, and probably the rest of the world. Their experience, knowledge and continuous service to the industry extending over 50 years, covers the whole period of the scientific study of lubrication problems, and of the realization that oils can be derived, for the service of industry, from all three kingdoms – animal vegetable and mineral. Three large factories at Glasgow, Liverpool and London were occupied with the refining and blending of oils. These were under the direction of an administrative staff composed of engineers and chemists. The chemists associated with the company were recognized as leading authorities in petroleum technology. The laboratories played an important part in the activities of the London, Liverpool, Glasgow and New York branches and large sums of money were devoted to research into problems connected with the internal combustion engine, Diesel, turbines and transformers. For example, in the grid electrification of Great Britain the transformers and switchgear supplied by the leading electrical manufacturers called for several millions of gallons of insulating oil of which more than fifty percent was supplied by W. B. Dick. The largest transformer in the world, at the time (125,000 hp) being manufactured by the Hackbridge Electric Construction Co., for the County of London Electric Supply Co., Ltd., was filled with Dick's Ilo Non-sludging Transformer Oil.

It was reported that in the early premises of the founder (*William Bruce Dick*) Mr. James Young, his first chemist, was the means, by his discovery of mineral oil from shale, of establishing the Scottish mineral oil industry. It was also believed that in these premises Mr. Young first produced paraffin oil, so the firm of W. B. Dick and Co. began early its tradition of research in this industry; the House of Dick occupied a unique position in the history of lubricants. Mr. W. Bruce Dick's slogan, as a well-known oarsman in his day, which was also that of his brother Mr. Charles E. Dick, the well-known international golfer, was "There is always room at the top for the best".

> The company had established depots for reserves of oil in many of the leading ports of UK, Europe, New York, New Zealand, Australia, Singapore, Shanghai and South America.

It is highly unlikely that James 'Paraffin' Young could have been an employee of W. B. Dick since he had founded his own company to exploit the many processes which he had developed and patented, from as early as 1841. He patented the process of distilling paraffin from shale oil and discussed his discovery at a meeting of the British Association in Edinburgh in 1850. He went on to form a partnership with Mr. Binney and Mr. Meldrum, establishing a large works at Bathgate, for the manufacture of Young's paraffin oil. Young's British patents expired in 1864, so it is much more likely that Dick, along with many other competitors, 'all apparently prosperous', exploited the opportunity to benefit from Young's early work. In the 1850's Young was involved in many legal actions to protect his rights. He is credited as the founder of the Scottish shale-oil industry which at its peak supplied 90 percent of the world's shale oil.

In later years W. B. Dick and Company, Ltd. became a major supplier of lubricants to the aircraft industry. Flight Magazine, on January 31st 1930, reported –

> Another firm to see the value of aircraft to their business is W. B. Dick & Co., of Grosvenor Gardens, whose chairman has just bought a Coupé Gipsy Moth, which will be flown by Mr. E. C. Brown and we expect to see him at the flying meetings which are shortly beginning again. This firm is one of the oldest in the lubricating oil business and has for many years acted as refiners to some of the best known lubricating oil companies, besides supplying lubricating oil to all Government Departments. Now the company are going to market their own aero engine oil under the name of ILO and the Moth will be used for a practical test of some 100 hrs flying. Both a mineral and castor oil will be marketed, but the former, in common with the growing tendency, will be the one generally advised for the use of private owners under normal conditions. Oils suitable for lubricating aero engines have, of course, been marketed in the past, but a somewhat extensive campaign is now being initiated and the purchase of this Moth is part of this campaign, and the engine will be run entirely on their own oil, its behaviour should prove a very good selling point.

In 1945 W. B. Dick and Co. were acquired by C. C. Wakefield and Co.

From November 1883 the Railway side of W. B. Dick & Co. had become Dick, Kerr & Co. railway & tramway contractors, and by 1885 had moved to Kilmarnock. Their first large contract was the Edinburgh Northern Cable Tramway, which Dick, Kerr part-financed.

At some date subsequent to this, the Dick, Kerr head office was moved from Kilmarnock to 94 George Street, Edinburgh, and John Kerr set up house at Luffness House, Drem; *it is speculated that the move was probably due to Dick, Kerr & Co. gaining a 21 year lease of the Edinburgh tramway.*

Kerr, Stuart & Co

From 1883 the careers of James and John Kerr follow separate paths. In 1883 James Kerr founded Kerr, Stuart & Co., at 135 Buchanan-street, Glasgow, in partnership with John Stuart. Their business was that of agents for locomotive manufacturers, including Falcon, John Fowler and Hartley, Arnoux & Fanning; also builders of railway and tramway plant. During the previous ten years James Kerr had invented several improvements in connection with portable narrow gauge railway plant. In 1884/85 Kerr, Stuart designed and completed seventy miles of light-railway for the Morvi Light Railway, India, and in 1885 supplied three locomotives manufactured by Dick, Kerr & Co., Kilmarnock.

James Kerr failed to benefit by the future success of the company he created, dying on 4th Dec. 1884, age 33. Earlier that year he had become a Member of the Institution of Mechanical Engineers.

The Glasgow Trades Directory for 1890 gives Kerr, Stuart, & Co. at 135 Buchanan-street, and a trade advertisement of the time has Kerr, Stuart & Co. of 20 Bucklersbury, London, E.C. and at Glasgow, makers of Portable Railway and Rolling Stock, Tramways and Cars for Steam, Horse, or Electric Power; Light Passenger & Goods Railways, Bridges, Roofs, &c.

In 1892, Kerr, Stuart bought Hartley, Arnoux & Fanning and moved into the California Works in Stoke to begin building their own locomotives. That part of the Hartley, Arnoux & Fanning business which was involved with the building of railway and tramway plant was sold to Dick, Kerr and Co.

Hartley, Arnoux & Fanning, originally Hartley and Arnoux Brothers, were in the business of manufacturing machinery for the pottery and brick making industries, they also undertook millwright and general engineering work. In 1886 Walter H. Fanning joined the firm, which was renamed, Hartley, Arnoux & Fanning; the company being described as railway and potter's engineers.

Locomotive building and other railway work was carried out on behalf of Kerr, Stuart and Company of London and other firms, and the first record of work carried out for Kerr, Stuart & Company appears as a drawing of a tipping box wagon dated 12th October 1888. The first locomotive built to the order of Dick, Kerr was named John Bull and its destination probably St. Lucia, British West Indies and despatched on 8th March 1891 (*reference: British Steam Locomotive Builders, James W. Lowe*).

Kerr, Stuart were known for producing a number of standard designs with many engines being built for stock and sold 'off the shelf' to customers. The names of these locomotive types were often derived from the purchaser of the first of that type or from the name it was given. Kerr, Stuart specialised in industrial locomotives both for standard and narrow gauge, more particularly the latter for home and abroad. Later main line types were built and some 1500 in all, up to the year 1930.

In 1894, Kerr, Stuart & Co., was incorporated as a private limited company.

In 1900, Kerr, Stuart designed a siege train locomotive, supplied to the War Office, for use in South Africa. The Engineer for 19th January 1900 reports –

> 'Siege Train for the War'; a very smart piece of work was done in connection with this train. The inquiry reached the London office of Kerr, Stuart and Co. on Monday 13th November. Practically nothing required was actually complete at the date of the inquiry. The order for two narrow-gauge locomotives, five miles of straight and one mile of curved railway, with steel sleepers &c., thirty sets of points and crossings, two sets of diamond crossover road, twenty-four four-wheeled wagons to carry 3½ tons each, fifteen bogie wagons to carry six tons each, and two six ton bogie brake wagons, was given on Saturday 18th November to the company's head office in London, and the majority of the work, including both locomotives was inspected and passed by the War-Office on the Wednesday following, 22nd November, four days later, including the Sunday.

Kerr, Stuart had a large joiners shop and a significant passenger coach construction business. They were, therefore, well placed to build the steam railcar. Their first railcar, built in 1904, was a diminutive 2 ft gauge saloon for the Maharajah of Gwalior. This was followed, in 1905, by a batch of 11 standard gauge railcars, six for the Taff Vale Railway, two for the Lancashire and Yorkshire Railway, two for the Great Western Railway and one for the Great Indian Peninsula Railway. The largest rail motor order was for 15 from the Italian State Railways. Over the next 25 years the California Works continued to build locomotives to their own designs and to the customers' own designs. In addition to British railway companies their list of customers included: Ireland, Chile, Mexico, Madras and Shanghai-Nanking. Following the First World War, and in common with most British locomotive builders they received a number of large orders from the mainline companies who were seeking to replace obsolete inherited equipment with their own standard designs. Customers included: the Metropolitan Railway (1920), London, Midland and Scottish Railway (1925 - 1927) and Great Western Railway (1929/30). They also exported to other countries including India and Brazil. In the late 1920s they produced a number of diesel locomotives with some success but development stopped when Kerr, Stuart went into receivership. The Register of Defunct Companies shows Kerr Stuart & Co. Ltd. registered in 1910, a court order being issued on 14th October 1930, released by the Liquidator on 4th May 1932 and struck off the register on 20th March 1936. The firm's goodwill (designs, spare parts, etc) was bought by the Hunslet Engine Co., locomotives to Kerr, Stuart design continued to be built by Hunslet also by W. G. Bagnall who had recruited their Chief Draughtsman and a number of other staff.

Kerr, Stuart had a working agreement with the Leeds firm of Robert Hudson & Co. who were manufacturers of light railway equipment and this relationship was maintained by the Hunslet Engine Co.

A particularly noteworthy employee of Kerr, Stuart and Co. was Reginald Joseph Mitchell who after leaving Hanley High School at the age of 16 gained an apprenticeship with Kerr, Stuart & Company at their Fenton, locomotive engineering works in the centre of Stoke-on-Trent, and the firm provided him with 'a sound basic training in engineering'. Reg, apparently, found this 'not entirely to his liking'. All apprentices had to start work in the engine sheds, and this meant that he had to get up very early and travel down to Stoke on the workmen's tramcar, to make matters worse, he had to wear overalls and carry his own lunch and tea can. At night he had to return home with grimy hands and with his overalls plastered with oil and dirt. When he complained to his father he was told "You will go my lad and like it". At the end of his apprenticeship he worked in the drawing office at Kerr, Stuart and attended the Wedgwood Burslem Technical School taking night classes in engineering drawing, higher mathematics and mechanics, since he had already decided to make a career in engineering.

In 1916, at the age of 21 he left Kerr, Stuart and made two attempts to join the Forces, each time being turned down for reason that his engineering skills would be of more use in civilian life. Whilst looking for work he did some part-time teaching in Fenton Technical School.

In 1917, he applied for a job as personal assistant to Hubert Scott-Payne at the Supermarine Aviation Works at Woolston, Southampton and was offered an interview followed by an offer of employment. In 1919 he became their Chief Designer; in 1920 he was appointed Chief Engineer, then in 1927 Technical Director, eventually achieving fame as the designer of the Spitfire fighter.

Another well known employee who was an apprentice at the works in the period prior to closure was Tom Rolt (L. T. C. Rolt), engineering historian and an early campaigner for the preservation of the canal network. He was the author of 'Narrow Boat' and many other works covering the field of transport heritage.

DICK, KERR & COMPANY

Britannia Engineering Works, Kilmarnock

Kilmarnock in the 1870s was a centre of engineering and locomotive building, one pioneer being Andrew Barclay who in 1847 acquired land from the Duke of Portland and erected the first part of the Caledonia Works. In 1856 the Glasgow & South Western Railway constructed a new works in Kilmarnock and the first locomotive, also the first to be built in Kilmarnock, was turned out in 1857. Andrew Barclay's business prospered with the development of the coalfield and iron industries and in 1859 he commenced the construction of locomotives at the Caledonia Works. This aspect of the business expanded rapidly and by 1870 had turned out about 100 locomotives.

In 1872 Barclay's general manager Allan Andrews, decided to join some local businessmen and set up a rival establishment. In June 1872 they purchased land from the Duke of Portland and erected the Britannia Works and in October a partnership consisting of Thomas Barbour, Duncan Gilmour, William Ramsay, Robert Anderson Jr, William Rankin Jr and John Yeudall with Allan Andrews took over the premises. The new venture traded under the title of Allan Andrews & Co. and opened for business in February 1873. The firm undertook all kinds of engineering work, and although locomotives do not seem to have formed a significant aspect of the business, the expertise of Andrews in that field ensured that some attention would be devoted to them; the first one was built for the G&SWR and despatched from the works on Tuesday 30th September 1873. At about this time Allan Andrews, with some of his partners, decided to invest in the Cumberland iron trade, the partnership being formed on 1st August 1873 traded under the title of the Derwent Hematite Iron Co., and Andrews went to Workington to supervise the erection of their works. Soon after the new company started operations most of the partners in the company gave up or retired from the partnership to be replaced by John Barr of Paisley and David Graeme Morrison. It would appear that an 'antagonistic attitude' soon developed between the two new partners and the two remaining partners from the original partnership; the firm now traded under the title of Andrews, Barr & Co. By 1880 a depression in trade was taking its toll on Kilmarnock and Andrews was having problems within the partnership and most of his business interests were experiencing financial difficulties, including loans that could not be repaid, and by early 1881 the business was operating under the supervision of creditors and Andrews was forced from the business. By 1882 the business was trading under the title of Barr Morrison & Co., and during their period of ownership at least one 0-4-2ST locomotive was built for overseas and it has been estimated that 35 to 40 locomotives were built altogether. The saddle tanks were of all shapes including the 'ogee' pattern, straight sided and semi-circular.

In January 1884 Barr Morrison & Co. were declared bankrupt with liabilities of £11,700 and assets of only £2,814-12s-3p and in May 1884 the trustees sold the works to Dick, Kerr & Co., the estate yielded a dividend of 4s in the pound. Dick, Kerr & Co. continued with locomotive building up to 1918. Before this date Dick, Kerr had no works of its own, the orders being sub-contracted. D. G. Morrison remained with Dick, Kerr until 1886, and the company had the use of Morrison's patents, including one for a tram engine. Many steam trams were supplied for home and abroad and they were the sole makers of Morrison & Kerr's patent tramway engines. It has been estimated that over 50 steam locomotives were built over this period.

The range of products was increased, and included rolling stock, track-work, roofing, bridges, small steamboats and 'every kind of plant and machinery', such as -

Axles, blowing engines, boilers, cable tramways, capstans, colliery tubs, cranes, crossings, gas engines, mining equipment, pithead gear, points, portable track, pumping engines, punts (for sugar cane plantations), sluice gates, stationary steam engines, tipping wagons, tram engines, traversers, turntables, wagons, water tanks, weighing machines (for cattle), wheels and winding engines, etc.

Their contracts included, Deptford Creek Bascule Bridge, Coatbridge Generating Station, Las Palmas Test Tanks, a 90 km. Railway in Portuguese East Africa and the first Staines Reservoir, etc. About 300 men were based at Kilmarnock in 1890, with the works occupying a four-acre site.

The portable railway was a patent of John Kerr's. It consisted of narrow gauge track with 10 lb. or 24 lb. rail, in sections that could be carried by a few men, with steel sleepers and rail joints that clipped together. It was designed for industrial applications, such as brick pits, plantations and sugar estates, and in places where the track would need to be lifted and moved frequently, usually by unskilled labour.

When Hartley, Arnoux & Fanning were taken over by Kerr, Stuart, the railway and tramway plant section was bought by Dick, Kerr & Co. and Mr. Hartley became manager at Kilmarnock.

From 1900 onwards the business of the Kilmarnock works appears to be that of a small scale manufacturer of steam locomotives, up to 1918; contract work on tracks and cable ducts for new tramway systems and supplier of mechanical equipments to the Electric Railway and Tramway Carriage Works, later the United Electric Car Company at Preston. The works were sold in 1919 to the Kilmarnock Engineering Co. and the works of Dick, Kerr and Co. Ltd. concentrated at Preston as part of the newly formed English Electric Co.

Steam Locomotives

The construction of steam locomotives and trams continued at the Britannia Works, after the take-over by Dick, Kerr & Co., up to 1918. The locomotive output of the Dick, Kerr works at Kilmarnock can only be thought of as small scale compared with other manufacturers. Total output between 1884 and 1918 was only about 106, including 18 classed as steam trams, an average of about three a year. They nevertheless catered for an international market, overseas customers taking approximately ⅓ of the total production output. Their range of customers including –

YEAR(S)	CUSTOMER	TYPE	GAUGE
1885	Alford & Sutton tramway	0-4-0(Tram)	2ft-6in
	Kerr, Stuart & Co for the Morvi Light Railway, India	0-4-0ST?	2ft-6in
	Mersin, Tarsus & Adana Railway, Turkey	0-6-0T	4ft-8½
C1885	Kerr, Stuart & Co for Penang Tramways, Malaya	0-4-0 (Tram)	Metre
1885/86	Kerr, Stuart & Co for sugar plantation, Brazil	0-4-2ST	narrow
1886	West Carbery Tramway & Light Railway Co Ltd, Ireland	0-4-0 (Tram)	3ft
1886/87	North London Tramways	0-4-0 (Tram)	4ft-8½
1890	Borgo San Donnino-Salsomaggiore Tramway, Italy	0-6-0T	4ft-8½
	Calcutta, India	0-4-0 (Tram)	
1891	Farleigh Sugar Mill, Queensland, Australia	0-4-0IST	2ft
1893	Cadbury Brothers Ltd, Bournville	0-4-0ST	4ft-8½
1896	Kure Vavel Factories, Japan	0-4-0T	
1896	Marian Sugar Mill, Queensland, Australia	0-4-2T	2ft
1901	Boyd & Forrest, Contractors, Kilmarnock	0-4-0ST	4ft-8½
	Namaqua Copper Co Ltd, Namaqualand, South Africa	0-4-0ST	2ft-6in
1902/07	Pahang Consolidated, Sungei Lembang, Malaya	0-4-0ST/0-4-2ST	2ft
1903	Manchester Corporation, Electricity Department	0-4-0ST	4ft-8½
1904/05	Cape Copper Co Ltd, Namaqualand, South Africa	0-4-2IST/0-4-2T	2ft-6in
1905/05	British North Borneo, Exploration Syndicate Co Ltd	0-4-2T/0-4-0T	Metre & 2ft
1905	Marconi's Wireless Telegraph Co Ltd, Clifden, Ireland	0-4-0ST	2ft
1907	Burma Corporation Ltd, Burma Mines Railway	0-4-2T	2ft
1908/10	Karachi Port Trust, Karachi	0-4-2ST	2ft-6in
1911	Yeni Light Railway Co Ltd, Burma		75cm
1912	Admiralty, Chatham Dockyard, Kent	0-4-0ST	1ft-6in
1914	Carrara Versalia Electric Railway & Power Co Ltd, Italy	0-4-0ST	Metre
1914/16	Taquah & Abosso Gold Mining Co Ltd, Gold Coast	0-4-2IST	2ft
1915	Admiralty, Chattenden & Upnor Railway, Kent	0-6-2T	2ft-6in
	Dick, Kerr & Co Ltd, Littleton Reservoir contract	0-6-0T	4ft-8½
1915/16	Ministry of Munitions, Hackney	0-6-0T	4ft-8½
1918	Islip Iron Co Ltd, Islip Quarries, Northants	0-4-0T	3ft

Key to locomotive type – T = Side Tank, ST = Saddle Tank, IST = Inverted Saddle Tank.

The list is by no means complete; James W. Lowe (British Steam Locomotive Builders 1975), Russell Wear (The Locomotive Builders of Kilmarnock 1977) and the Industrial Railway Locomotive Journal, published by the Industrial Railway Society, provide a much more comprehensive listing with supporting information and further technical data.

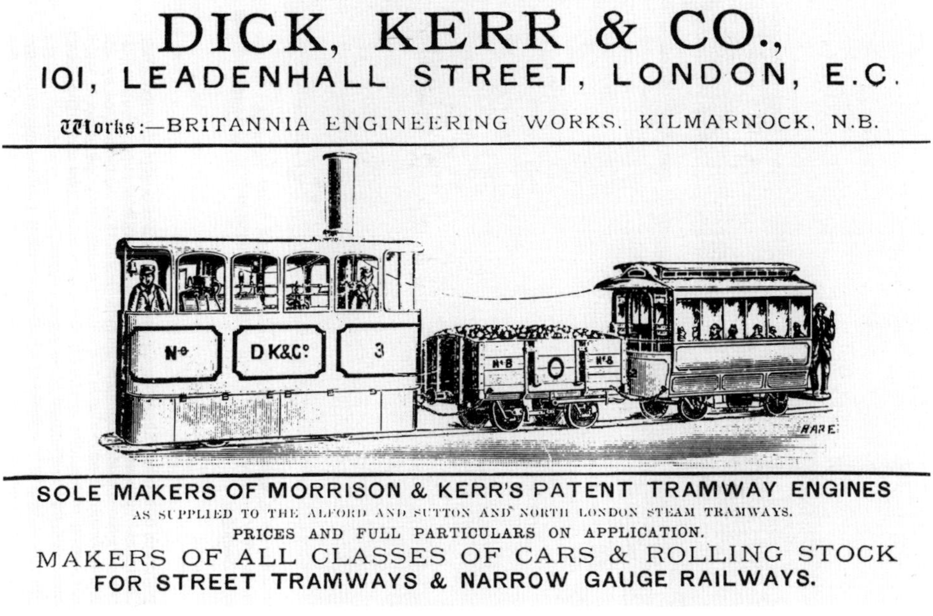

Advertisements for Dick, Kerr Steam Trams. Britannia Engineering Works, Kilmarnock, c1886.

Advertisement for Dick, Kerr Products c1886.

DICK, KERR & Co.
LIMITED
ENGINEERS
AND
CONTRACTORS
LONDON
AND
KILMARNOCK 1901

Northern Railways Locomotive MTR No. 2. Exhibited at the National Rail Museum, Shanti Path, Chanakyapuri, New Delhi. Manufactured in 1910 by, Dick, Kerr & Co. Limited at the Britannia Engineering Works, Kilmarnock. "A local shunting and freight engine, originally worked at the port of Karachi for 7 years and then used at the Marala Timber Depot before ending up on the Northern Railway at a Creosote Plant at Dilwan". Renovated by the Amritsar Workshops.

The Britannia Engineering Works, Kilmarnock and a Range of Narrow Gauge Rolling Stock.

Dick, Kerr & Co. of Kilmarnock built a tramway engine in 1885 for the Alford & Sutton Tramway, Lincolnshire. It had a locomotive boiler and inside cylinders of 7 inch bore and 12 inch stroke; wheels 2ft. 4 in. diameter and base 4ft. 6 in; the exhaust steam was superheated in the firebox (Morrison's patent). A similar engine was sent to Penang Tramways, a short lived concern that in 1894, having got into financial difficulties, came into the possession of Kerr, Stuart & Co. who formed a limited company, Kerr, Stuarts Penang Steam Tramways Ltd. In 1886-7 10 engines were built for the North London Tramways. Except for the design of the iron cab they were almost identical to the Falcon design, with two 8 inch by 14 inch inside cylinders, fitted with Morton's patent radial valve gear; wheels, 2 ft. 6 in.; wheelbase, 5 ft.; weight, about 10 tons. Three similar engines were despatched to Calcutta in 1890

Schull & Skibbereen narrow gauge railway

In 1886 Dick, Kerr supplied three 3ft gauge, 15-ton 0-4-0T steam trams to the West Carbery Tramways & Light Railways Co. Ltd, which from 1892 was renamed the Schull & Skibbereen Tramway and Light Railway Co. This 15 mile stretch of line, in a remote region of West Cork, was opened in 1886 and rapidly demonstrated most of the things that could go wrong. Due to a shortage of funding the building work had been 'skimped', with light rail and excessive curves and gradients (as steep as 1 in 22); to make matters worse small and underpowered but cheap locomotives were provided, which were burdened with an unsuccessful experimental boiler design and constantly broke down. The situation only improved, after two years of this, when the directors were replaced by a Committee of Management appointed by the local authorities, but only after the investment of much more money and the purchase of much more powerful, but still quite small, 24 ton 4-4-0T locomotives, after which they had a useful train service. Of the three steam trams supplied by Dick, Kerr, Marion was scrapped in 1906, Ilen in 1914, having been rebuilt following the initial fiasco, and Ida in 1925.

Trams in London

The North London Tramways Act 1883 sanctioned the use of steam on the lines operated by The North London and Suburban Tramways Company Ltd for a period of seven years, a heavier section rail being used to carry the increased weight of steam trams. The early tram engines were purchased from Merryweather & Sons, followed in 1886-7 by ten trams from Dick, Kerr & Co. Ltd., of Kilmarnock, these were numbered 16 to 25 in the company fleet list.

Dick, Kerr Steam Tram, Fleet Number 18. One of 10 Dick, Kerr engines put into service on the Edmonton Line by the North London Tramways Company in 1886 – 87 in readiness for the extension to Wood Green.

Namaqua Copper Company

A 7 Ton locomotive Pioneer was acquired by the Namaqua Copper Company in 1901; described as a plantation type 0-4-0ST of 2ft 6in gauge, *'Pioneer'* was too light to use on the main line from Braakpits Junction and was first used on the tramway to Tweefontein. Pioneer's main claim to fame came on 1st May 1902 when it derailed 'without much damage', outside O'okiep, on the Cape Copper Company mainline, after being 'kidnapped' by the Boers of Concordia (the town having surrendered to the Boers on 4th April 1902) and used as a mobile bomb in an attempt to explode a wagon load of dynamite in O'okiep. *Pioneer* pushed the wagon, which was loaded with 8 tons of dynamite towards O'okiep which had been protected by a barbed wire fence across the railway line at Braakpits Junction. When the wagon breached the fence it derailed at a set of points located just inside the fence. The dynamite was thrown to the ground and ignited but did not explode, burning out harmlessly.

Steam Launches

The Glasgow Herald for 23rd July 1885 reports under the heading 'A novel industry in Kilmarnock' -

> Dick, Kerr and Co., locomotive engineers and ship owners with headquarters in London have commenced to construct, for themselves, six steam launches in the Britannia Works at Kilmarnock which have hitherto been used exclusively for engineering purposes.

The Griffin Gas Engine

In 1888 Messrs. Dick, Kerr, and Company of London produced, under Griffin's patent, the Griffin Gas Engine, reported in some detail in The Engineer journal for 18th May 1888, then the Glasgow Herald for 28th August 1888 –

> Steadiness of motion, with a special view to use in electric lighting, has been attained in the 'Griffin' engine, several specimens of which are exhibited by Dick, Kerr & Co., Kilmarnock. The engine in its general style and method of working more closely resembles the ordinary steam engine than any other forms of gas engine. This engine has recently been tested by Professors Kennedy and Jamison and both their reports were highly satisfactory. *Professor A. B. Kennedy, M.I.C.E., visited the Kilmarnock Works on several occasions, at the invitation of the Dick, Kerr, Co., and carried out complete tests of several types of Griffin's patent gas engines which they constructed*. The engine was found to work very steadily, there being practical uniformity of speed at full power, no pulsation or blinking being observable in the incandescent lamps.

Advertisement for the Griffin Gas Engine. Produced by Dick, Kerr and Company, Kilmarnock in 1888.

> The mean speed was 223.8 revolutions per minute, and the gas used per indicated horse-power per hour during the trial was 18.92 cubic feet and was practically the same in each of the two-hour periods, which is a first-rate result.

Also in 1888 Dick, Kerr & Co exhibited the Griffin gas engine at the Great Centennial Inter-colonial Exhibition in Moor Park, Sydney, Australia, during Centennial week, under the auspices of the Agricultural Society of New South Wales.

An advertisement of August 1890 –

> The 'Griffin' a perfect gas engine, double action accomplished, increased power, no false explosions, simple, steady, uniformly superior; awarded a Gold Medal for general excellence by the Society of Arts. The makers were Dick, Kerr & Co. of London and Kilmarnock.

Blowing Engine

The Glasgow Herald for 31st January 1889 reports –

> Messrs. Dick, Kerr & Co. of London and Glasgow, and the Britannia Engineering Works, Kilmarnock, have secured the contract for a large blowing engine from Messrs. Charles Cammell & Co., for the Workington Furnaces.

This will be the fourth set of engines of this class that the Britannia Works have built for Messrs. Cammel & Co.

The above provides some indication of the range of products coming from the Kilmarnock Works during the early years; this, however, represented only a relatively small part of the company's overall business. Over the same period Dick, Kerr and Co. were becoming increasingly active and acquiring a good reputation in the business of tramway construction.

Early Tramways

Before 1860, tramway development was confined mainly to the United States and to a lesser degree to mainland Europe. The modern tramway was introduced to England by George Francis Train an American entrepreneur, who, in 1857, made proposals for laying tramways on the system originated in Philadelphia in some of the Metropolitan thoroughfares and in a few provincial towns. Mr. Train in association with James Samuel, C.E. failed in a bid to obtain an Act of Parliament, which was applied for in 1858.

In March 1860, Train applied for, and in May of that year obtained permission from the Commissioners of Birkenhead to lay down his tramway in their town. In April he patented his system. On 30th August 1860 Train opened Britain's first street tramway running between the Woodside Ferry terminal and Birkenhead Park, Birkenhead, a distance of 1½ miles.

'With the permission of the local authorities' lines were also laid down in London, by Train and 'his friends'. During 1861 in Bayswater Road, between the Marble Arch and Notting Hill Gate; in Westminster, from the Palace Hotel to Victoria Station; and in Kennington Road, from Westminster Bridge to Kennington Park. In 1863 a line 1¾ miles in length was laid in the Potteries District for the Staffordshire Potteries Street Railway Company, between Burslem and Hanley.

The interest in tramways which was spurred on by these early experiments made it necessary for the Government to consider legislation to reduce the amount of Parliamentary time required to process all the applications. The first Tramways Act went onto the statute books on 9th August 1870 and provided protection for those local authorities through whom the schemes were promoted. The Act attempted to promote the new means of transport by clarifying and regulating the legal position. It authorised local boroughs or urban district councils to grant a 21 year concession to a private tramway operator. The operator could construct the track as part of the concession but was responsible for the repair of the public highway between the tracks and a short distance either side. The local authority could construct the track themselves if they wished to retain complete control of the highway, but they were not allowed to operate trams.

At the conclusion of the lease, the local authority could purchase the complete undertaking, including the trams and horses (or, in the case of a steam tramway, the locomotives and trailers).

Several sections of the Act were later repealed or superseded by other legislation, including the Light Railways Act 1896. Most notably, local authorities were given the right to construct and operate their own tramways under the newer Acts, and municipal ownership became the norm when the original concessions expired.

The passing of these acts resulted, over a very short time, in a flood of applications from tramway entrepreneurs in all parts of the country and particularly in the London area. Acts of Parliament approving these Bills came thick and fast, and eventually, most sizeable towns and cities, together with many of the smaller towns had horse tramways approved and built. Like the situation on the railways a couple of decades earlier, the impact of tramways on the streets of the towns and cities of Britain would be considerable.

Horse Tramways

John Kerr had been active in street tramway construction even before the formation of Dick, Kerr & Co. In 1880 he devised an improved form of track for horse and steam tramways. Known as John Kerr's first system, this consisted of girder rails made from Bessemer steel weighing 58 lb per yard and rolled in lengths of 24 feet. These were laid on continuous wrought iron sleepers of inverted trough section, bedded in concrete to a gauge of 3ft 6 inches. A full scale trial was necessary, and William Bruce Dick took up a large shareholding in the newly formed Ipswich Tramway Company of 1881, which was laid using Kerr's first system.

In 1879 the Board of Trade had granted permission for the construction of a horse tramway in Ipswich. Horse trams had become popular as the tram rails provided a smoother surface to travel over compared with the bumpy roads of that period. The line opened for public service on 13th October 1880, serviced by two single deck tramcars each pulled by a single horse. Ipswich Corporation ran the horse trams until 6th June 1903 after which the network was converted for electric traction.

In 1902 Dick, Kerr and Co. installed 10.8 route miles of track at 3ft 6in gauge. The first trial run of the electric tramway was on 10th November 1903 with an official opening the following day; public use started on 23rd November.

Kerr's second system used girder rails, 4½ inches high, weighing 45 lbs. per yard, laid to a gauge of 2½ feet on transverse concrete sleepers spaced 3 feet apart. For rail fastening, steel tie-plates, ¼ inch thick, were laid on and embedded in the sleepers. This system was laid (as a demonstration) on the Alford and Sutton steam tramway, using 12½ ton steam engines.

A Horse Tram of the Edinburgh and District Tramways Co. Ltd. A Subsidiary of Dick, Kerr & Co., Incorporated in March 1894.

Kerr's third system was laid with girder rails, 5 inches high, weighing 60 lbs. per yard and laid on timber sleepers, 8 inches by 4 inches deep, and spaced 9 feet apart. This system was first used in Bucharest and Madrid.

At that time the structural testing of steel girder rails for tramways involved subjecting a short piece of rail, 5 feet long, placed on supports 3 feet apart and subjecting it to a blow from a weight of 1 ton falling freely on its centre point, from a given height, without causing more than 1 inch deflection at the centre. The rail would be subjected to two drop tests, and for example, a rail of 60 lbs. per yard as used with Kerr's third system, would be from a height of 6 feet and then the same piece of rail from a height of 15 feet without showing any signs of fracture. A test was then carried out to assess rail deflection and recovery when subjected to a static load for a period of thirty minutes. On a specimen of rail of 98 lbs per yard, as used for the Huddersfield Corporation Tramways, a dead load of 25 tons was applied at the centre of the specimen and when the load was removed there was no permanent set.

The Harrow Road & Paddington Tramways Company

The Harrow Road & Paddington Tramways Company was incorporated as a statutory company by the Harrow Road & Paddington Tramways Act of 1886; the company was empowered to construct a line from the Royal Oak at Harlesden Green, along the Harrow Road to a point west of the Lock Hospital Bridge over the Grand Junction Canal at Amberley Road in Paddington. The line involved 2.33 miles of double and single track on the standard gauge and ran from a point about 0.162 miles from the Amberley Road terminus with branches along Chippenham Road, Malvern Road and Cambridge Road. At the time these lines were promoted the area served lay wholly in Middlesex, but most of the company's lines were in areas which passed to the newly-formed London County Council on 21st March 1889.

The company was incorporated with a share capital of £60,000 in £10 shares, and the first directors were John Kerr, Frederick Manuelle, John Metcalfe, Benjamin Nowell and Joseph Robson. At a later date new directors were appointed and some of them were also directors and shareholders of the North Metropolitan Tramways Company.

The construction of the line was completed by July 1888, Maj. Gen. C. S. Hutchinson carried out the Board of Trade Inspection on 2nd July and the line was opened for public service on Saturday 7th July. Initially the car fleet consisted of twelve two-horse double deck cars and 110 horses. In 1889 the company deposited plans for an extension of the system over a distance of 1 mile, 1 furlong, 3.75 chains from the junction of Amberley Road, to the eastern end of the Harrow Road, from there a distance of 1 mile, 2 furlongs, 8.5 chains along Edgware Road, Maida Vale and High Road, Kilburn.

The Harrow Road and Paddington Tramways Bill debated in the Commons on 12th March 1903 'To empower the Harrow Road and Paddington Tramways Company to construct new tramways; to work their tramways by mechanical power; to lease their undertaking to the Metropolitan Electric Tramways, Limited; and for other purposes', was read the first time and referred to the Examiner of Petitions for Private Bills (Hansard, 12 March 1903). The lines became part of the Metropolitan Electric Tramways system in 1904.

River Hospitals Tramway and Joyce Green Hospital, Dartford

In the latter part of the 19th century, local government in London was concerned about fever and smallpox cases among the poor and in 1870/1 opened hospitals in Hampstead, Stockwell and Homerton. Complaints about such hospitals in built-up areas resulted in a Royal Commission being set up and its recommendation was that smallpox should be treated in isolated situations on the banks of the Thames or in floating hospitals on the river itself. Two old wooden battleships were acquired and anchored off Greenwich; in 1884 they were moved to a location off Dartford Marshes. In 1893 the Board decided that with the risk of fire or collision always possible the ships should be replaced by a permanent smallpox hospital on shore. A site was chosen at Joyce Green but construction was delayed until 1899. However, work did go ahead with the provision of a horse tramway along a causeway from Long Reach Pier to the site for the new hospital to ensure a smooth journey for the patients.

The contract for the roads and tramway went to Mr. J. Dickson of St. Albans who sub-contracted some of the tramway work to Dick, Kerr & Co. The tramway was a single line of 4 foot gauge with passing places and was completed by May 1897.

The River Hospitals tramway was not used until the smallpox epidemic of 1902. Five purpose built ambulance trams were acquired – one large ambulance car in 1905 from the Electric Railway and Tramway Carriage Works and in 1909 two large and two small ambulance cars from the United Electric Car Company of Preston.

Cable Tramways

Britain's first cable-operated street tramway was opened on Highgate Hill in North London in 1883 by a subsidiary of the Patent Cable Tramway Corporation Ltd., the UK agent for the Hallidie patents, under the supervision of Messrs. Eppelsheimer and Colam, the line was opened in May 1884. This company went into liquidation in 1887 and the Dick, Kerr Co. bought the UK patents from the liquidators, thus ensuring that they would have the first option to participate in any further cable tramways in Britain. In May 1884 the Patent Cable Tramways Corporation had acquired ownership of the City of Bath horse tramway system when the Bath Tramways Co. Ltd. ran into financial difficulties. The new company went into liquidation in 1888 and on 11th August 1888 the Patent Cable Tramways Co. sold out to Dick, Kerr & Co. who the following year sold the system to the Bath Road Car & Tramways Co. Ltd. The horse trams continued in Bath until July 1902, it would be another 6 years before the city had an electric tramway system.

The early tramways were built before the Dick, Kerr Company had its own car-building capability, resulting in the cars being sub-contracted to G. F. Milnes, Falcon, Brown Marshall or Metropolitan. Dick, Kerr built the tracks, the gripping gear, the winding engines and other machinery, placing sub-contracts for the boilers, buildings and cables. Further opportunities came along, firstly in Edinburgh-

The Edinburgh Cable Tramway

It would be impossible to write a complete account of the History of the Dick, Kerr Company, and particularly the career of John Kerr without, at least, a summary of the planning, construction and further development of the Edinburgh cable tramways. From about 1884, and whilst he was still a young man of 31, this project seems to have been of particular personal interest to John Kerr, and probably a major reason for making Edinburgh his home town. His first, and unsuccessful, attempt to win a seat in Parliament was in the General Election of 1900 in his home constituency of Haddingtonshire. After he was successfully elected to represent Preston, Lancashire in the by-election of 1903 his sole contribution (ref; Hansard for 15th June 1904) was on the subject of the Leith Corporation Tramways Order Confirmation Bill on which he was unable to speak due to his "being pecuniarily interested".

Dick, Kerr and Company Advertisement from 1894.

The Edinburgh Street Tramways Company was incorporated on 29th June 1871 by the Edinburgh Tramways Act of 1871, which authorised the company to construct specific tramways 'to be worked by animal power only' in Edinburgh, Leith and Portobello.

The Edinburgh Northern Tramways Company was formed on 7th August 1884, for the construction of, what would be, the first part of the Edinburgh Cable Tramway system to replace the horse drawn trams; the choice of cable traction was driven by the steep gradients on many streets in Edinburgh.

The Patent Cable Tramways Corporation Ltd. were the contractors and although work started in 1885, it was October 1886 before track laying started, along with the erection of an engine house in Henderson Row. The Engineer, for the contractor, was a 'young man' (aged 32) by the name of William Newby Colam who had been Eppelsheimer's Chief Assistant at Highgate. The opening of the first 1½ mile section of the route took place on 28th January 1888, without ceremony. The system was found to work well and early in 1889 work started on the 1½ mile Stockbridge route which started service in February 1890. The engine house was provided with two non-condensing horizontal steam engines of 300 hp manufactured by Dick, Kerr and Co., Kilmarnock; Messrs Babcock and Wilcox provided the boilers.

In 1887 Dick, Kerr & Co. had acquired from the liquidators of the Cable Tramway Corporation Ltd. the English patents of a revolutionary tramway system introduced in San Francisco by Andrew Smith Hallidie, a British engineer.

Andrew Smith Hallidie was born in London on 16th March 1836 to Andrew Smith, a Scottish schoolmaster, and Julia (nee Johnstone). In 1852 he with his father sailed for California, where his father had acquired an interest in some gold mines in Mariposa County; the venture was unsuccessful and in 1853 the father returned to England. Andrew remained in California and became a gold miner, blacksmith, surveyor and builder of bridges.

Andrew had adopted the surname of Hallidie in honour of his uncle, Sir Andrew Hallidie, who had been a royal physician to William IV and to Queen Victoria.

In 1856 whilst working on the construction of a flume at a mine he was consulted over the excessive wear on the ropes being used to lower cars loaded with rock from the mine. His solution was to improvise a machine for making wire rope which had a working life of ten times that of the old ropes. In 1857 he set up a works in California for the manufacture of wire ropes, under the name of A. S. Hallidie & Co. and in 1857 he invented and patented the Hallidie ropeway, a form of aerial tramway for transporting ore in mountainous districts by means of an elevated endless travelling line. In 1871 Hallidie completed plans by which street cars could be propelled by underground cables and in 1872 Hallidie and his associates, A. E. Hovay, W. Eppelsheimer and Henry Root constructed a cable tramway in San Francisco from the intersection of Clay and Kearny Streets to the crest of the hill, a distance of 2,800 ft, making a rise of 307 ft. The tramway proved to be a success and was in use by the public from 1st September 1873 following a satisfactory demonstration on 1st August. The Hallidie patent was employed around the world wherever the gradients could not be catered for by conventional trams, including Edinburgh, and he became a very rich man.

Tramcar for the Edinburgh Cable Tramway Car Number 209 Built in 1903 by the Electric Railway and Tramway Carriage Works, Preston and photographed at the Works.

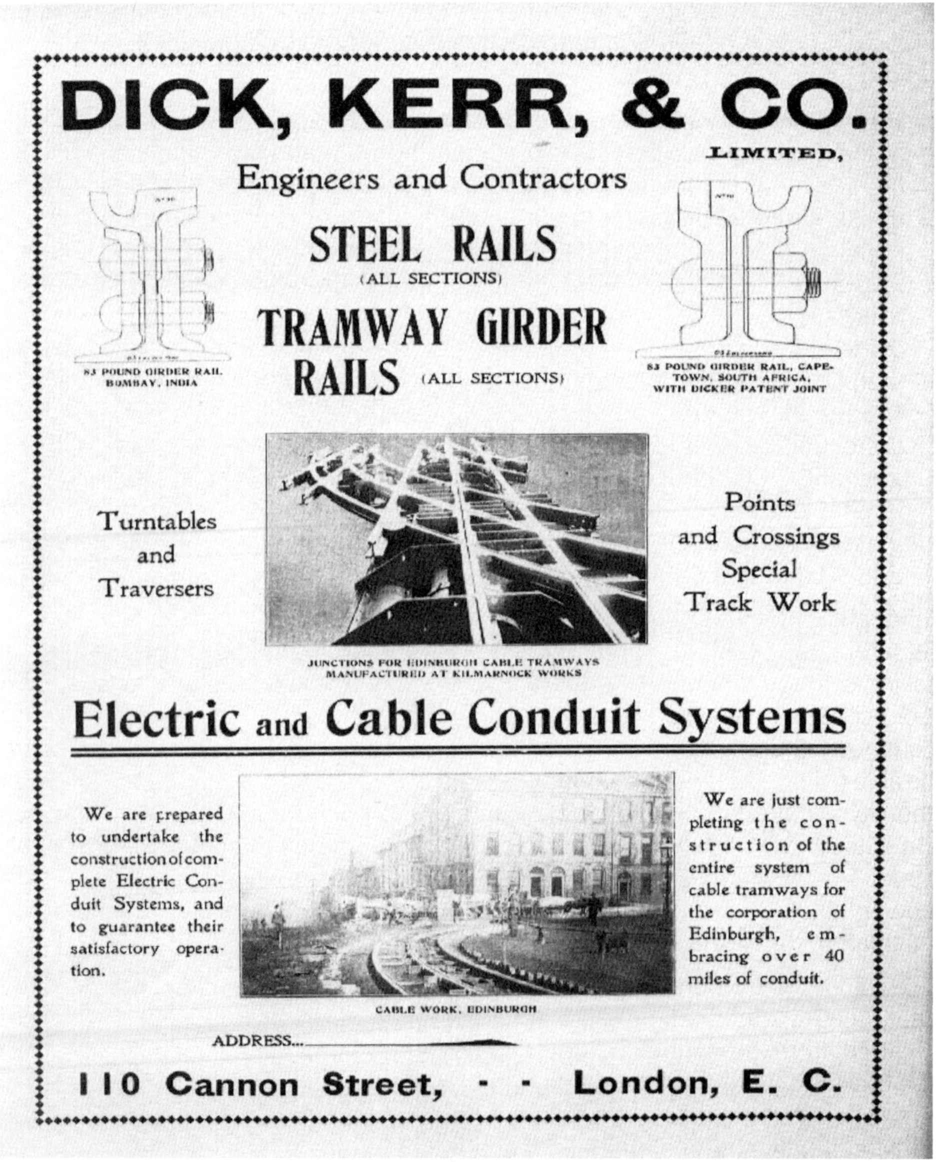

Dick, Kerr and Company Limited Advertisement from 1899.

Reported in the Engineer Journal for 21st February 1890 was a short report on the opening of the new line of cable tramways between Princes Street, Edinburgh and Stockbridge; the route taken by the tramway is described street by street –

> The cable used for operating the new line was worked by the same engine which, at the company's depot in Henderson-row, has been hauling the cable cars on the Trinity route since January 1888. The depot was a considerable distance from the Stockbridge route, and the cable had to be taken along Henderson-row and Hamilton-place in a tube before it joined the main route. Owing to the peculiar situation of the engine-house to the Stockbridge route, it was necessary to take the main cable across the uphill track before diverting it into the main track tubes, and consequently an auxiliary cable was introduced for conveying the cars round the sharp curve from the main cable in Hamilton-place to the main cable in Kerr-street. The auxiliary cable received its motion from the main cable, and it was set in motion or stopped by the conductor of the car. The speed of the main cable was initially six miles an hour. The outside seats of the cars were reversible to enable passengers to face the direction of travel. The line was constructed from the designs and under the superintendence of Mr. W. Newby Colam, A.M.I.C.E, by Messrs. Dick, Kerr, and Co., of London and Kilmarnock, who were represented by Mr. James More, C.E.

In August 1890, with the Edinburgh Street Tramways Company's first 21 year lease drawing to a close, Edinburgh Town Council announced their intention of taking over the system within the city. At the same time Leith and Portobello Town Councils entered into an agreement with the Edinburgh Street Tramways Co. for their continued operation of their parts of the system. The company hoped they would be able to lease the Edinburgh part of the system from Edinburgh Corporation when the city acquired it, as local authorities still had no power to operate tramways. On 12th August 1892 Edinburgh Corporation gave statutory notice to the Edinburgh Street Tramways Co. and to the Board of Trade, of their intention to exercise their powers of acquisition in terms of section 43 of the General Tramways Act of 1870, of that part of the system within the city boundaries.

The Edinburgh Corporation Tramways Act, passed on 29th June 1893 authorised the use of cable power and offers to lease the system were sought. Three offers were received, from the Edinburgh Street Tramways Co., Messrs Dick, Kerr & Co. Ltd., and the Glasgow Tramways Co. At a council meeting on 11th October 1893 Dick, Kerr & Co's offer was accepted, and it was arranged that they should enter on the property on Monday 20th November.

On 9th December 1893 the Corporation took possession of the Edinburgh Street Tramways Co's system within the city, amounting to 11¾ miles of tramway, about 60 cars and some 600 horses at the valuation of £40,000 and immediately resold to Messrs Dick, Kerr & Co. Ltd., who took over the working for twenty one years from that date.

The Edinburgh and District Tramways were formed as a subsidiary of Dick, Kerr & Co. and on 9th December 1893 it took over the Edinburgh Street Tramway Services within the Edinburgh city boundary.

Shortly after Dick, Kerr & Co. Ltd. concluded their lease of the Edinburgh system they approached the Corporation for permission to sub-lease the undertaking to a new company which they proposed to form. This was with a view to raising additional capital for their Edinburgh activities, and they deemed it sound to attract this capital locally and invite some prominent local gentlemen to join the Board. In July 1894, after a considerable amount of debate, the Corporation agreed to the existing lease being determined and entered into a new lease jointly with Messrs Dick, Kerr & Co. Ltd., and their new subsidiary company, the Edinburgh & District Tramways Co. Ltd., which had been incorporated on 6th March 1894. The first directors were B. Hall Blyth (chairman), Alex Fleming, G. H. Geddes, John Kerr, and Frederick Manuelle. On 9th March 1894 and reported by the Edinburgh Evening News –

> Among the new joint stock companies registered in Edinburgh during the past week is the company bearing the title 'The Edinburgh and District Tramway Company (Limited)'. The company proposes to enter into and carry into effect an agreement, dated 9th February 1894, between Dick, Kerr & Co. on the one part, and Alex George on behalf of the company on the other, and to acquire when practicable, the right to work the tramways in the said agreement, or other tramways within a radius of 10 miles from the Register House, Edinburgh. The nominal capital of the company is £117,500, and the subscribers include Messrs H. Hall Blyth, John Kerr, Alex Fleming, S.S.C., R. H. Johnston, W.S., and J. S. Tait, C.A.
>
> From July 1894 the rolling stock carried the name of the Edinburgh & District Tramways Co. Ltd., in place of Dick, Kerr & Co. Ltd; Dick Kerr & Co. Ltd. remained liable to the Corporation for the financial affairs.

Meanwhile debate continued within the Corporation on the proposed introduction of cable traction and it was not until March 1895 that a decision was reached, and in July they appointed Mr. W. N. Colam and Mr. John Cooper as Engineers; the latter being Burgh Engineer at the time. Tenders were invited and in March 1896 that of Dick, Kerr & Co. Ltd. was accepted, from fourteen offers, to carry out the work of converting about 23.75 miles of tramway, including the 3 miles of original cable tramway. William Newby Colam became engineer for the company, apparently representing both the constructor and the client, although this was refuted in the local press. Extensions to the tramway system were provided for by the Edinburgh Improvement and Tramways Act, passed in August 1896 which included the line to Meadowbank and on 1st January 1897 it took over the Edinburgh Northern Tramways cable operated lines.

The Edinburgh cable tramways inaugural ceremony took place on Wednesday 9th September 1896 –

> On the invitation of Messrs Dick, Kerr & Co., the contractors for the laying of the cable tramways in Edinburgh, the members of the Town Council and a large number of friends assembled at Craigmillar Park for the purpose of seeing the breaking of the ground by Lord Provost Macdonald in connection with the new system of cable tramways for the city; Messrs Dick, Kerr & Co. the Corporation Lessees were also the contractors for the laying down of the line. Mr John Kerr on behalf of Messrs Dick, Kerr & Co. presented the Lord Provost with a 'beautifully ornamented' spade in solid silver, with old oak shaft, and bearing a suitable inscription, and invited his Lordship to break the ground. In his speech the Lord Provost "regretted that they had to work the traffic for so many years by means of horse power, but now they had reached a period of importance, and he had no doubt that the constructors would not only do their work speedily, but he hoped they would do it well". At the inaugural luncheon John Kerr took the chair and presented the Lord Provost, whose health he proposed, with a 'fine silver salver, bearing a suitable inscription'.

By the early part of 1899 the Council and the public, 'were getting impatient for the cable cars'. The Tramways Committee met regularly to hear Mr. Colam's progress report; fourteen cars had been delivered, over a hundred more were ready at the maker's works –

> On 1st June 1899 the Tramways Committee, the Lord Provost, Mr. Pitcairn (the manager), Mr. Colam, Mr. John Kerr and Mr. More of Messrs Dick, Kerr & Co. and several other notabilities assembled at Shrubhill at 12.30 where they boarded one of the new cars, number 142, promptly on the sound of Edinburgh's one o'clock gun, the car moved off for the inaugural trip. The system was not yet available for public service, a considerable amount of work remaining to be done. Eventually the system was ready for inspection and on 12th October Sir Francis Marindin, the Board of Trade Inspector, made his inspection. On 19th October the Town Council held their 'official Inspection' and public service commenced on 26th October.

On 20th October 1899 the Engineer reported on the examination of the first section to be completed of the Edinburgh Cable Tramway system over the route between the city boundary at Pilrig Street, Leith Walk, and the suburban terminus near the Braid Hills; the length of the track involved was approximately 8½ miles –

> Accompanying Sir F. Marindin on his tour of inspection, there were Sir Andrew McDonald, Lord Provost of Edinburgh, and several members of the Corporation; Mr. Hall Blyth. C.E: Mr. Pitcairn, general manager of the Edinburgh and District Tramway Company, who are the lessees for the cable as well as horse tramways in the city, Mr. John Kerr, representing Messrs. Dick, Kerr and Co., the contractors; Mr. Colam and Mr. Cooper, the city engineers; and Mr. Arthur J. Shackleford, of Brown, Marshall and Co. Limited, Birmingham, the makers of the Cars. The article describes the construction and working of the Tramroad in some detail. Sir Francis Marindin had to leave soon after the inspection but at a luncheon afterwards,

> given in the Royal Hotel, Sir Andrew McDonald was able to report that "he was very pleased with everything he had seen and that so far as he was concerned, he absolutely passed for traffic the line they had gone over".
>
> On 12th July 1901 there was a further report on the 'Cabling the Edinburgh Tramways'.
>
> This large and important undertaking, commenced over four years ago, is now approaching completion, and as numerous improvements and novelties have been introduced in the construction of the lines, it will give to Edinburgh the best system of cable tramways in the world. The system includes 42.4 miles of track, extending over 21 miles of street, and with three exceptions, viz., those of Kansas City and San Francisco with their 60 miles of track each, and Melbourne with its 88 miles, forms also the greatest in the world worked by one company. Edinburgh is a city which, from its hilly characteristics and the consequent steep gradients which have to be traversed, is peculiarly suited for cable traction, as the weight of the descending cars assists materially in dragging the ascending cars up-hill, and effects a saving in steam power which no other system of haulage could bring about. The work presented a good many difficulties on account of the numerous crossings, junctions, and curves which had to be dealt with, and the inventive genius of Mr. Colam, the Chief Engineer, has been called into requisition at every step. --- One distinguishing feature is the number of cables wrought from each power station, as many as five being worked from one station.

The article goes into considerable detail, with particular attention given to the design and construction of the track, the rail joints and the cable grippers. The fish plate used for the joining of two sections of rail being described as –

> Presenting a departure from the usual practice and on one side is carried up flush with the tread to take the place of the part of the rail cut away, and thus forms half of the bearing surface for the wheels. There is no break through the entire breadth of the tread, but only breaks extending halfway across in three different places. The arrangement produces a very firm and smooth joint, and has been patented by Dick, Kerr and Co.

At the beginning of 1908 Dick, Kerr & Company was extending the Edinburgh cable system in two districts of the city. The Gilmore Place route replaced a section of line that was still operated by horse drawn trams. The second extension was a new route in the Broughton district.

Brixton Road, London (from contemporary accounts)

By 1870 a horse tram route ran from Kennington Park, then up Brixton Hill to Water Lane. The route was run by the London Tramways Company. When a further extension to Streatham was proposed, the problem arose of how to counter the steep Brixton Hill; one solution considered was to build a cable tramway. The Brixton Cable

Tramway was built to a gauge of 4 feet 8½ inches; and ran from near Kennington Park through the busy thoroughfares of Brixton, up to Streatham Hill. The depot, situated at Streatham terminus, was built to accommodate the machinery and cars necessary for working a proposed extension of the line to Streatham Common, the extension was never built but the line was extended to Streatham Library in 1895.

This system was designed by W. Newby Colam for the London Tramways Company, and executed by Messrs. Dick, Kerr & Company, and their engineer, Mr. More who were contractors for the road, gripping gear, engines, and most of the machinery. Messrs. Babcock & Wilcox supplied the boilers; Messrs. Cradock & Co. the cable; and Messrs. Lucas & Aird the buildings. It was opened at the beginning of 1893, and was 2¾ miles long, but an extension was operating from January 1896, bringing the total length to 3½ miles, and thus, at the time, forming the longest cable road in the country. The style of construction was similar to that in Edinburgh.

The Street Railway Journal for January 1896 reports on the completion of the new extension to the Brixton (London) cable tramway –

> The street construction is of the same type as that designed by Mr. W. N. Colam for the older section of the line, and his original plans of the power station provided for this extension. As, however, it had been contemplated that the new portion when built should be single line, and as, through a widening of the street, it has been found possible to construct a double track, there was a good deal of difficulty in radically altering as well as enlarging the pit under the tracks in front of the power house without interfering with the running of the cars on the old section. The contractors, Messrs. Dick, Kerr & Co. and their engineer, Mr. James More, Jr., were successful in carrying out this work without any interference with the car traffic. The tramway was, on the day, the longest cable road in Britain, being three and a half miles in length, or seven miles reckoned by single track. Two cables are employed.

Matlock: Bank Road

The Matlock Cable Tramway was a remarkable piece of engineering for its day, enabling both residents and visitors to get up and down Matlock Bank with ease. The single line tramway from Crown Square, near the bridge, to Matlock Bank, constructed with the assistance of Sir George Newnes, Bart., under the provision of the Tramways Act, 1870, was opened on 28th March 1893.

The overall length of the track was about one mile with the steepest gradient up Matlock Bank, covering a distance of about half a mile and an altitude of 300 feet with a maximum gradient of approximately 1 in 5 and claimed at the time to be 'probably the steepest in the world'.

The Matlock Cable Tramway Constructed by Dick, Kerr and Co. of Kilmarnock and opened in March 1893. A tramcar on the 1 in 5 gradient.

The local newspapers of the time reported the opening in some detail –

> The inauguration of the new tramway was in every respect a satisfactory one. Everything combined to make the proceedings unusually successful and the auspicious event will, it is hoped, be an augury of great success, commercially and otherwise, for the new venture which was yesterday opened by Mr Newnes. That the tramway will considerably enhance the attractions of the beautiful Derbyshire inland health resort is unquestionable. Visitors will not have passed the meridian of years, and who are sound in wind and limb, are not likely to find fault at being compelled to climb the lofty and breezy heights, which are one of the charms of the district, but for invalids and those advanced in age the limited facilities of cheap locomotion has been a most serious objection. However, to overcome this difficulty has been a question which has occupied the attention of many prominent and worthy residents during recent years, and it is probable that the matter would still have been deferred but for the intervention of Mr. George Newnes, M.P., who supplied the necessary capital.
>
> – Somewhere about the middle of 1890, after he had had a number of consultations with his local coadjutors, Mr. Newnes instructed Mr. Marks, C.E., of Birmingham, to prepare [the Parliamentary] plans for the work on a gradient of 1 in 5½, which makes the line the steepest in the world. The road up Matlock Bank was so narrow that only a single line could be run, but this difficulty has been met by what is a novel system – a single cable line, which serves alike for the up and down track for the same cars to run upon, with a passing place at the two main roads at Smedley-street. The cable itself is of Craddock's best steel, having 42 wires and being three inches in circumference, stranded onto a patent system for resisting wear and tear. It travels at 5½ miles per hour, and it is supported in a concealed culvert by a number of specially designed pulleys arranged some on each side of the culvert to carry the two portions of the cables, which are travelling in opposite directions. Each car possesses a gripper to embrace the moving cable. The gripper has one moving jaw or block, which, upon being raised by the controlling hand wheel, causes the cable to be seized and the car thus to be carried forward by it; and when the wheel is turned so as to open the jaws the cars remain stationary while the cable travels on, the car being checked by a brake.
>
> Mr. G. Croydon Marks, of Birmingham was the engineer of the line; Mr. Edward Marks was the resident engineer and Mr. Colam the consultant engineer. The whole of the machinery and the line equipments was supplied and laid by Messrs. Dick, Kerr and Co., of London and Kilmarnock.

The tramway service ended on Friday 23rd September 1927 when the cable broke. The bus service that was introduced to replace the trams on 1st October 1927 was unable to make the journey up the hill and the bus companies found alternative routes to get round the problem of the steepness of Bank Road.

Upper Douglas Cable Car

The tramway was the result of a public demand for some means of communication between the Promenade along the shore and Upper Douglas, which was seen to be a necessity if the upper part of the town were to enjoy the prosperity and growth to which it might reasonably look forward. Such importance was attached to the construction of a tramway through the upper streets that when early in 1895 the Tramways Company came to the Douglas Town Commissioners to ask for a renewal of the lease of the horse tramway along the promenade the Commissioners expressly stipulated, as an indispensable feature of the agreement, that the upper part of the town should be directly connected with the landing pier and the promenade by a tramway. With a gradient of 1 in 10.6 it was determined that a cable system would be the only practicable solution. A deputation of the officials representing the tramways company, the Douglas Town Board, and two members of the corporation, visited London and inspected the Brixton and Streatham cable tramway of the London Tramways Company. They then conferred with Messrs. Dick, Kerr and Company, Limited, the contractors for the London tramway, and obtained particulars with reference to the cable tramways also constructed by that company in Edinburgh and Matlock. The deputation was then relieved of any doubt as to the suitability of the cable system for working the proposed route in Douglas. As a result of the investigations the directors of the Isle of Man Tramways Company subsequently entered into a contract with Messrs. Dick, Kerr and Company, Limited, for the construction of the tramway.

In July 1895 the Upper Douglas Tramway Bill was put before Tynwald and in November 1895 the Act was passed.

> The construction of the track and conduit resembled closely the cable tramways in London and Edinburgh. The track, of 3 ft. gauge, consisted of steel rails of the grooved girder type and weighing 76 lb. per yard. Both the cable conduit and the foundation of the track were formed of Portland cement concrete; the conduit being 19 in. deep by 9 in. wide. The cast-iron yokes which supported the slot rails were firmly placed on a foundation of solid concrete, and were spaced 3 ft. 6 in. apart. The slot rails were bolted to the yokes, leaving an aperture three-fourths of an inch in width for the gripper shank to pass through. The weight of the slot rail was 38 lbs. per yard. On the straight lengths of track vertical pulleys 12 in. in diameter, with grooved rims, were placed at frequent intervals whilst horizontal pulleys, located on curves, had flat faces and broad rims. The spacing of the horizontal pulleys was dependent on the radii of the curves.
>
> In addition to the pulleys which guided the cable in the conduit there were large diverting pulleys, 8 ft. in diameter, in the main pulley pit outside the depot and also at the terminal pits. The entire line being worked by a single cable passing from a tension carriage along the tension race and subway to the diverting pulley

about which it is directed then along supporting and guiding pulleys to the terminal pit. Here it was led around two vertical pulleys of 8 ft. diameter, then passing the engine house and descending to the terminus. From there it was drawn up the hill to the main pit where it was diverted into the subway and taken around the driving pulleys and back to the tension carriage, this arrangement was considered to be comparatively simple. The cable was formed of the finest steel wire laid on a hemp core, with six strands of thirteen wires each, seven round six. It was 20 tons in weight and entirely in one length.

A power station was erected which included boiler and engine rooms, reel house, car shed, smithy, workshop and offices. In the boiler room were two Lancashire boilers, with Galloway tubes. Two engines, of 250 horse-power each were of the horizontal high pressure non-condensing type, with high pressure cylinders 20 in. in diameter and 42 in. stroke and fitted with the latest patent automatic variable releasing gear and spring governor to control the output of the engine to meet the varying loads on the cable. The cylinders were steam jacketed with live steam, having drain pipes and automatic traps. The flywheels were 13 ft. 4 in. in diameter.

The whole of the works, including engines, driving gear, conduits and depot buildings were constructed and carried out by Messrs. Dick, Kerr and Company, Limited and their chief engineer (*James More*) was responsible for the designs.

On 6th August 1895 the threading of the cable was completed, and on the following day the first car was drawn over a part of the line by the cable. The official inspection was made on 9th August by James Walker, C.E., Harbour Board Engineer; the inspector, suggested a slight alteration in the special slot brake to serve some of the curves and, after alarm bells were attached to the cars, a certificate was granted, and the line was opened for traffic on 15th August. Gripper men from London and Edinburgh were sent to Douglas to instruct the local men in the use of the gripper and brake mechanism, and in a few days the 'probationers' became fully experienced, and were allowed to manipulate the cars without assistance.

On Thursday 27th August 1896 Messrs. Dick, Kerr, and Company gave a dinner at the Douglas Bay Hotel to celebrate the opening of the Upper Douglas Cable Tramway; a large number of local gentlemen and others interested in the enterprise being present.

The first twelve cars were supplied by G. F. Milnes of Birkenhead, who up to this date were the exclusive supplier of Manx rolling stock. Subsequent to this, in 1907 two cars were supplied by the United Electric Car Co., then in 1909 and again in 1911 two cars from Milnes Voss bringing the total fleet to 16 cars. The tramway ceased operations on 19th August 1929.

A Registered Company

By 1890, and co-incident with the construction of the Edinburgh cable tramway, Dick, Kerr and company having become firmly established in the business of tramway construction were in need of capital to expand, John Kerr found the financial backing that was needed and took the necessary steps to register a new company under the same name.

Dick, Kerr, and Company, Limited was registered on 31st May 1890 and reported in the business pages of the local and national press on 2nd June with details of the prospectus. The Engineer on 13th June, under the heading of New Companies, reports –

Dick, Kerr, and Company, Limited

This company was registered on the 31st ult., with a capital of £160,000, in £5 shares, one-half of which are 7 per cent, cumulative preference shares, to carry on business as manufacturers of railway and tramway appliances, and as iron and steel founders and electricians. An unregistered agreement of 31st ult., between F. Manuelle and John Kerr, of the first part; G. Flett, of the second part; and G. W. Carr – for the company – of the third part, will be adopted. Messrs. Manuelle and Kerr are the promoters, and one of them is to be managing director. As consideration for the purchase of a business not specified in the articles of association, they are to receive £115,000 in cash and shares. The subscribers are -

F. Manuelle, 131, Leadenhall-street, merchant.
J. Kerr, 101, Leadenhall-street, engineer.
G. Flett, 13, Cavendish-road, S.W., engineer.
J. W. Carr, 54, Norray-road, Putney, cashiers.
A. M. Allison, 20, Lebanon-gardens, Wandsworth, clerk.
J. A. J. Smith, 22, Relf-road, Peckham Rye, clerk.
J. Leigh, 52, Thurlow-hill, West Dulwich, clerk.

The number of directors was not to be less than two, nor more than seven, the first were Major General C. E. Webber, C.B., C. Thornton Cayley, F. Manuelle, and J. Kerr; qualification £500 in shares. The remuneration of the directors would be £200 per annum in respect of each ordinary director, with £100 additional for the chairman. The board were also entitled to 10 per cent of all profits remaining after payment of the preference dividend, and providing for the reserve fund, and for 7 per cent dividend on the ordinary shares, but the total remuneration was not to exceed £5,000 per annum. John Kerr was appointed managing director for five years, at a salary of £1,200 per annum. The Company Solicitors were Messrs. Linklater and Co., 2, Bond-court, Walbrook.

Claud Thornton Cayley was appointed chairman.

The Liverpool Mercury -

The advancing years of the senior partner (*William Bruce Dick*), and the need for considerable extensions of the works and plant necessitated by the increasing business, have induced the firm to convert their enterprise into a joint stock company, whilst continuing to retain a substantial interest in the business.

The firm (the prospectus states) is old-established, and has an acknowledged reputation for the following specialities:- Supply and manufacture of tramway permanent way material, including the road construction; portable railway and rolling stock; supply and manufacture of railway permanent way material; railway points and crossings, turntables, tanks, &c; cable tramway machinery, and the construction of cable tramways; gas engines.

The Morning Post -

This company is formed for the purpose of carrying on, under enlarged and improved memorandum and articles of association, a business identical with that of the Municipal Trust Company, which was formed in 1879, and the company's affairs will be under the same management as the old company.

In 1893-1894 the company is listed in the General Directory for Glasgow as –

- Dick, Kerr & Co., Limited, engineers and contractors, 3 Dundas Street, City.
- Dick, Kerr & Co., Limited, gas engine manufacturers, and rolling stock and general engineers, Britannia Engineering Works, Kilmarnock.

The Street Railway Journal –

In October 1895 and reported, under the heading 'Dick, Kerr & Company, Ltd' –

The name of Dick, Kerr & Company has for a long time been well known on both sides of the Atlantic as one of the most prominent contractors in the European street railway field, and as selling agents for the well known 'Phoenix' type of rail, manufactured by the Phoenix Iron & Steel Company, of Laar, near Ruhrort, Germany. The offices of this company are located in Leadenhall Street, London, and their works are at Kilmarnock, N. B. The Canadian trade of this company has been very large, and the company is actively pushing the sale of its goods in the United States.

The street railway business of Dick, Kerr & Company, although comprising only a small portion of all its work, is very extensive. The company has taken contracts for the entire equipment of a number of cable railways in England, which have proved very successful, and it has built a large number of cars of various types for street railway use.

--- Dick, Kerr and Company, who act as sole selling agents of the Phoenix rails for Great Britain and her Colonies and the United States, have been associated with the Phoenix Company over fifteen years and have done much to render the products of this company popular among street railway managers.

In June 1899, Dick, Kerr & Company secured the order for eighty cars with their complete electrical (Walker) equipment, for the Liverpool municipal tramways and for thirty car bodies for Sheffield. All the car bodies to be made by the Electric Railway & Tramway Carriage Works, of Preston and the Walker motors to be made in America, although later motors would be made by the new works of Dick, Kerr & Company, to be located in Preston. The new works, the English Electric Manufacturing Co. Ltd, would require a considerable capital investment.

Dick, Kerr & Company Ltd. went into voluntary liquidation on 21st October 1899 and was reconstructed as a company of the same name. Shareholders were entitled to 2 fully-paid ordinary shares of £1 in the new company for each ordinary share of £5 held or to 3 fully-paid preference shares of £5 held. Holders of debentures were entitled either to £103 in cash or in 4% debenture stock in the new company. The final meeting return was registered on 21st December 1900.

This reconstruction was probably necessary in order to finance the acquisition of the works on the west side of Strand Road in Preston, recently constructed by the Manchester consortium of business men who had styled themselves 'The Equipment Syndicate'. The new works was registered as the English Electric Manufacturing Co. Ltd., a subsidiary of Dick, Kerr & Co. Ltd of Kilmarnock and London and registered in December 1899.

On 23rd June 1900 Dick, Kerr, and Co., Limited combined with the English Electric Manufacturing Co., Limited and the Electric Railway and Tramway Carriage Works, Limited, exhibited at the Tramway and Light Railways Exhibition in the Agricultural Hall, Islington, London. The large combined exhibit included numerous examples of different sections of permanent way and other items of tramway equipment, and four fine tram cars manufactured and equipped by these companies. One of these was a bogie vestibule top-seat car, with a capacity of sixty four passengers, for heavy city service. Another would seat fifty five passengers, and was intended for high-speed summer suburban traffic. The further two were four-wheeled cars, one being single and the other double decked, and having seating capacities of twenty six and fifty six respectively. In addition to these there were, exhibited, a truck equipped with motors and air brakes, and various motors, controllers, motor driven air compressors &c.

In their annual report for the year ended 30th June 1902 Dick, Kerr & Co., Ltd., of London were reported to have made a 'very satisfactory showing'. In their report to shareholders, on 30th September the directors 'consider the result of the twelve months trading satisfactory', and state that 'the prospects for the current year show every indication of good results'.

> Since the last report important contracts have been carried out for the construction and equipment of several electric tramways in Great Britain and abroad, and these works have given satisfaction to the company's clients. The resolutions for acquiring the shares of the English Electric Manufacturing Company, Ltd, were duly passed, and the exchange of the shares will be carried out. Steps will be taken to acquire the few outstanding shares and arrangements will be made with the debenture holders of both companies at an early date.

On 23rd August 1902 the Street Railway Journal, reporting from London –

> Dick, Kerr & Co, Ltd. of London who have for some time been sole selling agents for and large shareholders in the English Electric Manufacturing Company, Ltd., of Preston; have now taken over the whole of the shares of that concern. The business of Dick, Kerr & Co. has been conducted as a private limited company so far, its shares being held almost entirely by members of the firm. The amalgamation has been effected by Messrs. Dick, Kerr & Co. exchanging £100,000 ordinary shares for the £200,000 in the English Electric Manufacturing Company, and they have exchanged £145,000 out of the £185,000 preference shares in the latter company. The debentures have not yet been dealt with, but it is intended to convert the Dick, Kerr & Co. 4 per cent debentures to 4½ per cent, the same as those of the other company, and to raise the preference dividend to 6 per cent all round. The capital of Dick, Kerr & Co. now figures at £840,000 and its profits for the last three years have averaged over £60,000 which is sufficient to pay the increased interest on the preference stock and debentures in the two concerns, and also give a 10 per cent dividend on the ordinary. For the current financial year the profits are estimated to be considerably in excess of previous years. The English Electric Manufacturing Company has also been earning substantial profits. It is stated that during the last eighteen months these have been £67,000.

The ordinary general meeting held on 28th September 1903 at the Cannon Street Hotel, London was chaired by Claud T. Cayley and amongst those present was William Bruce Dick –

> Mr Cayley was able to congratulate the meeting on the very prosperous condition of the company, and to announce that the company were able to pay a dividend of 15 per cent., and had increased the reserve fund by £70,000. The company had recently acquired a large electrical machinery works at Preston (*the English Electric Manufacturing Company Ltd.*), and their present contracts for permanent ways, cables, rolling stock, electrical equipment, generating plant, etc in Mandalay, Singapore, Hong Kong, Tokyo, South Africa, South America, Australia, Canada and elsewhere amount in round figures to about 2½ million pounds. Mr. Dick with reference to the above moved a vote of thanks to the staff in London, and the officials at the Preston works.

Reported at the ordinary general meeting of the shareholders of Dick, Kerr & Company, Ltd., held in October 1905, John Kerr M.P., presiding –

> The net profits to be dealt with were £85,007-8s and £836-10s-10d, better than 1904. The amount of reserve originally settled to be provided for was £150,000. That amount was made up last year, so that the liability had been discharged. This year it was proposed to take from the profits £23,388-16s-2d, and coupled with the sum of £16,611-3s-10d, previously reserved to no particular object, to open an account for "special reserve for extensions of manufactures or replacements of machinery". The chairman urged the meeting to remember that an "industrial concern like theirs should make a study of safeguarding itself in every direction. In these days of rapid change in machinery, it was wise to provide a fund in excess of the ordinary ample depreciation, so that they could, if necessity arose, provide either works extension or the most modern machinery without issuing fresh capital. He hoped to see the special reserve fund considerably augmented in the near future. He trusted they would support the board in their recommendation in setting aside this very considerable amount of £40,000. The board, who represented a very large proportion of ordinary stock, had done it merely to strengthen the position of the company". The investments stood at £109,354-7s-9d., as against £109,701-14s-10d. He thought "anyone would be fortunate if they could acquire them at that valuation". The amount in the bank was £90,609-12s-3d, against £55,151-19s-8d last year, and the amount carried forward was £42,387-16s-5d, against £39,922-11s-4d.
>
> Claud T. Cayley, deputy chairman, seconded the adoption of the report and accounts, the motion being agreed.
>
> A dividend of 6 per cent per annum on the preference share capital, and a dividend of 10 per cent on the ordinary share capital were declared, a balance of £42,387-16s-5d being carried forward, and £23,388-16s-2d being set aside to special reserve.

William Bruce Dick, the founder of the company, had died on 11th July 1905; his obituary recognised him as founder of 'the electrical engineering firm of Messrs Dick, Kerr, and Co.,' but surprisingly made no mention of W. B. Dick and Co. The published record of the shareholders meeting makes no reference to this particularly poignant and significant event in the history of the company.

ELECTRIFICATION

Electric power was first applied to a propelling vehicle in 1879. In that year visitors to the Berlin Exhibition were transported through the grounds in small carriages hauled by an electrically driven locomotive. In the latter part of the nineteenth century the progress of electric traction became more rapid, it spread quickly to surface railway lines, being used particularly for heavy suburban passenger traffic and to replace steam locomotives on underground sections of line. The use of electricity in transport began to show its true possibilities with Dick, Kerr and company Ltd., being among the early pioneers to exploit the benefits of electric traction supplying electric traction and generating equipments to a number of railway companies.

Railway Electrification

Liverpool Overhead Railway:

The Liverpool Overhead Railway Company was formed in 1888 and opened for traffic in March 1893, and the whole seven miles of track was in full operation by December 1896 –

> The third rail system with 500 volts dc was used, current being supplied from three substations drawing power from the public supply. This was one of the first elevated lines in the world operated by electricity, the total length of the line was about 7 miles, with double track and a third-rail feed and a special rail return. The trains were made up of two or three cars, the larger trains weighing about fifty five tons. With the old equipment the entire distance was run in 32 minutes including 16 stops or practically at a scheduled speed of 12½ miles per hour. This required a minimum of 14 trains running at five minute intervals to handle the traffic. The original rolling stock consisting of twenty-two two-car units, each consisting of two composite motor coaches, each equipped with two 60 hp motors and direct series-parallel control. The conditions of the line, with fifteen intermediate stations and an average of only 760 yards between stops, militated against high speeds, and the city tramways, then in the course of development, were able to provide a faster service.

The importance of increased acceleration on electric railways was well shown by developments on the Liverpool Overhead Railway by means of the new type of equipment supplied by Dick, Kerr & Co. Ltd., which permitted a faster service better able to compete with the tramway system.

In 1901 it was decided to experiment with new equipment in order to ascertain what acceleration could be obtained, and Dick, Kerr & Company, Ltd. entered into a guarantee to run the whole distance, including 16 stops of 12 seconds, in 20.9 minutes; in other words to improve the scheduled speed from 12½ miles to nearly 19 miles per hour.

The stock was replaced by three-coach units, each consisting of two motor coaches with an intermediate trailer. Each motor coach was equipped with two Dick, Kerr 100 hp totally enclosed motors with series-parallel control, both sets being operated from a controller at each end of the train by means of a bus line cable. The weight of the train, including 154 passengers, was 55 tons. A series of tests and trial runs were made and it was clearly demonstrated that the new train would make the journey easily in less than 20½ minutes.

Reported at the time –

> The stimulating results of the competition have been evidenced in Liverpool by the development of the electric tram system. Prior to the introduction of the cars the Overhead Railway Company, itself an electrical undertaking had a practical monopoly of the passenger traffic along the dock line, having previously 'knocked out' the horse omnibuses. The advent of the electric trams has now exercised so serious an effect upon the traffic of the Overhead Railway that the directors of the latter undertaking had to boldly face the situation. The consequence has been a remarkable acceleration of the train service, which was inaugurated yesterday by some trial runs and an inspection of the railway and its equipment generally. Sir William Forwood, the chairman and a numerous company of guests attended. The speed now attained is 19 miles an hour, including stoppages, which is stated to largely exceed the speed of numerous similar railways both in this country and in the United States. The new equipment has been furnished by Dick, Kerr & Co.

The Waterloo and City Railway:

In 1893 the problem of transporting city workers from the South side of the river to business centres on the North side was becoming acute, particularly at Waterloo Station. The success of the City and South London Railway, which had been electrified in 1890, indicated the way in which a solution could be found and in 1893 powers were obtained for building a railway under the river, connecting Waterloo with the Bank.

The railway was ready for operation by 1898 and was opened for public use on 8th August of that year.

The initial rolling stock was powered by 60 hp gearless motors supplied by Siemens Bros. Dynamo Works Limited. In 1899 these trains were supplemented by five single-unit motor coaches, each coach being equipped with two 60 hp nose suspended motors with single-reduction gear drive. These coaches and their equipments were built as complete units in the Dick, Kerr & Company Limited's works at Preston and continued in full operation for forty-one years.

The Lancashire and Yorkshire Railway:

By 1902 the Lancashire and Yorkshire Railway (L&YR) had identified problems with station handling capacity on the Liverpool to Southport line using its existing steam locomotives; the General Manager took the decision to consider electrification using multiple unit trains. These, by their rapid turn round and high acceleration, would enable the traffic to be increased without extra station accommodation and would it was hoped attract traffic to the railway by providing a better service than a tramway could offer.

Estimates for the work were obtained from four different firms, both for complete electrification to Southport, and for only seven miles out of Liverpool. The result of this was an order on 22nd October 1902 to electrify the Liverpool- Southport line at 600V dc using a live third rail and multiple unit trains. The contract for the entire work, except for the construction of the car bodies, was awarded to Dick, Kerr, & Co. Ltd at Preston.

The main power house was erected near Formby on the bank of the river Alt which would provide cooling water; sub-stations were to be erected at Birkdale, Seaforth and Sandhills, and another at Formby power station.

Work began on 8th March 1903 and by 29th July the three separate sub stations were completed; the power station foundation and building was erected and the high-tension cables completed by 26th November, and on 20th December the first engine and generator were ready to supply power to the line.

The power station at Formby was equipped with four 1,500 kW generators, driven by slow speed cross compound engines, and one 4,000 kW Willans, Dick, Kerr turbo-alternator set generating 7,500 volts, 3 phase, 25 cycles. Current was converted in four substations through Dick, Kerr 600 kW rotary convertor sets and supplied to the third rail at 600 volts d.c. The main engines were supplied with steam from sixteen hand fired Lancashire Boilers.

On 29th December it was possible to make an experimental run with the first train.

A total of twenty-three miles of double track route had been electrified in eighteen months at a cost of about £340,000. On 6th March 1904 a train was run from Crossens to Liverpool on a trial trip of twenty one miles, stated to be the longest non-stop run made by an electric train up to that time in this country. On Sunday 20th March a train ran from Liverpool to Southport in fourteen minutes at an average speed of 70 mph, and two days later a partial service for the public began.

The electrification of this section of the Lancashire and Yorkshire Railway was the first example in Great Britain of a complete changeover to electric traction of a section of main line railway. The line, in all eighty-three single track miles, runs from the Exchange Station, Liverpool, to Crossens, a few miles north of Southport, and to Town Green, serving Aintree *en route*.

For the opening of the electrification Dick, Kerr supplied forty motor-car sets, and subsequently ten more. The standard train consisting of a motor car at each end and two trailers between; the complete trailers, and the bodies for the motor cars, were built by the L&YR at their Horwich and Newton Heath Works.

In 1904 with the intention of running a through service between Southport and Dingle, the L&YR in conjunction with Dick, Kerr built a special motor car, No 1000, with accommodation for twenty first-class and fifty third-class passengers. In order to conform to weight restrictions it was built with aluminium panels and fittings, and weighed only twenty-two tons; subsequently eleven more were built. The through service operated up to the outbreak of the 1914-18 war; afterwards the cars were used on the Southport-Crossens shuttle service until withdrawn in 1946. In May 1906 an extension to Ormskirk made it necessary to increase the capacity of the Formby power station. Dick, Kerr in conjunction with Willans and Robinson, of Rugby, were awarded a contract for the supply of a 3,000 kW turbo-alternator.

Bury-Holcombe Brook:

Following the successful conversion of the Liverpool-Southport line to electric traction, the L&YR had in mind the development of electrification in the Manchester area based on Manchester and embracing Bury, Oldham and Rochdale.

Dick, Kerr approached the L&YR in January 1912 for permission to experiment with a high-voltage dc overhead system. The short branch to Holcombe Brook, forming an extension of the Manchester-Bury line was chosen and an agreement was drawn up by which Dick, Kerr were to supply all the equipment, carry out the work and run the trains at their own cost. The project was in the hands of Thorsten Von Zweigbergk, as Technical Director of Dick, Kerr.

The electric trains went into public service on 29th July 1913; power was taken from the local commercial supply to a sub-station, near Holcombe Brook, where a Dick, Kerr 525 kW motor generator supplied direct current to the overhead line at 3,500-3,600 V. Two motor coaches and two trailer coaches were also supplied by Dick, Kerr to form a pair of two-car trains which could be driven from either end. –

Each motor coach was fitted with four 250-h.p. nose suspended motors, connected permanently in pairs in parallel.

Each unit of motor coach and trailer was provided with a driving position at either end, the master controller of which operated control gear, consisting of electro-magnetic contactors with five series and three parallel notches. Control current was supplied at 120 volts from a small rotary transformer.

The Lancashire and Yorkshire Railway, Bury-Holcombe Brook Electrification. The electrification of the line and building of the sub-station was carried out by Dick, Kerr and Company in 1912 and opened for public service in July 1913. The trains operated on 3,600 volts dc. This was the first high voltage dc installation in the world.

Lancashire & Yorkshire Railway Manchester to Bury Electrification Turbo Generator Plant in Clifton Power Station 1915.

Manchester-Bury:

The success of the Liverpool-Southport electrification encouraged the Lancashire and Yorkshire Railway to take a further step in electrifying their suburban services. A partial service was opened on the Manchester-Bury line in 1915 but the development of a complete service was delayed by the First World War.

The railway decided on a 1,200-volt dc third rail supply for this line. Although the 3,500 volt system with overhead conductors, as used on the Holcombe Brook experimental line, had presented few problems the third rail system was considered preferable, and 1,200 volts was considered to be the 'maximum practical voltage for third rail collection'.

The side third rail was wood protected on both sides and on top and contact was established by a spring loaded steel shoe carried on a wooden shoe beam fixed to the axle boxes. This feature was considered to be 'unique in British railway operation'.

Power supply was obtained from a generating station at Clifton built specifically for the purpose. The station was equipped with two 5,000 kW, 6,600 volt, 25 cycle turbo-alternating sets, and the current was converted to 1,200 volts dc in two substations, each equipped with three 1,000 kW rotary converters. All the generator plant and substation equipment were supplied by Dick, Kerr & Company Ltd.

The rolling stock, made up of sixty motor coaches with associated trailers, was built by the L&YR at their Newton Heath works. The motors, of which there were four per coach, were of the 'standard self-ventilated type' designed by Dick, Kerr & Company but built by the L&YR.

The Great Northern Railway Proposal:

Dated 29th June 1903, Dick, Kerr & Co. Ltd. submitted a comprehensive proposal for a 'Scheme for the Substitution of Electricity for Steam Traction on the Suburban Lines of the Great Northern Railway Co.' The lines extending to New Barnet, Enfield, High Barnet, Alexandra Palace and Edgware Road had a total length of 28 miles. The report went into considerable technical and logistical detail including rolling stock, performance, time tables, permanent way, power stations, power transmission and transformation and general costs with detailed illustrations. There is no evidence that this proposal was pursued any further at the time but it does provide a clear indication of the confidence the company had in its own technical and commercial capabilities, and of the abilities of their technical staff.

Electric Tramways

The steam tram had only just started to replace the horse-drawn vehicle when the electric tramcar became a feasible and more environmentally friendly alternative, the steam tram being both noisy and dirty. The electrification of street tramways, generally, was delayed in many areas by resistance from the proprietors of existing horse tramways and other vested interests including, understandably, those whose livelihoods depended on the horse. Nevertheless by the closing decade of the century street electrification was becoming a reality and once the benefits had been clearly demonstrated its introduction rapidly spread throughout the country.

In the 21st Century it is difficult to imagine the enormous contribution made by the introduction of the electric tram to our towns and cities. The horse tram was a significant step in the right direction but the obvious problems of feed and stabling for the horses, grooming, breeding new stock, disposal of retired and deceased animals, and the need to be constantly clearing manure from the streets were a major incentive for a better solution to our transport problems. In dry weather and particularly in summer any mess left in the streets became a health hazard and in wet weather would make life difficult for pedestrians. A new horse would be put to work at five years of age and be worked for about four years. About ten or eleven horses would be required for each tramcar in service. A typical working day for a horse, operating in pairs, would have been about 3½ to 4 hours in harness, hauling a tramcar some 14 to 15 miles. The larger companies would also have their own horse hospitals with premises for the convalescing and resting of animals. The horse had always been the constant and faithful servant of man, for work and transport, but the pressures of urban expansion and increasing population in association with technological progress provided the necessary stimulus for change. The introduction of the steam locomotive in the early part of the 19th Century had started the revolution in land transport by providing comfortable and affordable cross country transport for the masses whilst the need for comparable street transport was being neglected.

It would be beyond the scope of this account of the company's history to consider all the contracts bid for, or entered into, by Dick, Kerr & Co., nevertheless those contracts which are considered to be of some significance from a historical, technical or business perspective, are described in some detail.

The history of tramway development in the United Kingdom and the difficulties experienced by those who had the foresight to promote their introduction and the tenacity to push for the necessary enabling legislation is briefly covered, if only to clarify its significance to the development and success of the Dick, Kerr Company.

Volk's Electric Railway:

The first electric tramway to provide a regular service in Great Britain, and one of the first electric lines in the world, was built by Magnus Volk, a Brighton born inventor, at Brighton in 1883; it was essentially a light railway and in its extended form remains in operation today.

This was a significant milestone in the early development of electric traction, and although W. B. Dick and John Kerr had no involvement with this project they would have known about it and in all probability have followed its progress with great interest and been aware of the possible implications with regard to their own future ambitions, since this event coincided with their take-over of the Britannia works at Kilmarnock.

Magnus Volk had shown an interest in things mechanical from his early years, and in his teens was always inventing things, trying out new ideas; he equipped his house with the first telephone and electric light system in Brighton. The Corporation then employed him to supervise the installation of electric light in the Royal Pavilion, which he completed in 1883. At about this time he made an electric motor and this together with a Siemens D.5 dynamo and a 2 hp Crossley gas engine he realised he could devise a system to move a railway car. He sought permission from the Corporation to build a short line on the sea front to try out his plan and a six month licence was granted; Magnus moved his gas engine and dynamo to the Royal Humane Society's Arch under the promenade. A Brighton man, J. T. Chappell, laid him a light 2ft gauge track on a shingle embankment extending about a quarter of a mile east from the Swimming Arch, to the old Chain Pier. The 50 Volt output of the power plant was fed to the two running rails; another Brighton man constructed a simple four wheeled car to which Volk fitted the motor using a belt to drive the axles, the current reaching the motor through the wheels via wires rubbing the tyres, and returning through the wooden centred wheels to the opposite running rail.

The opening ceremony took place at 12 noon on Saturday August 4th 1883, when the car carried the Mayor and other dignitaries as well as Magnus, and on the following Monday, a Bank Holiday the tram shuttled up and down the line for 11 hours and doing a 'great trade'.

In January 1884 he closed the line for reconstruction and extension. The gauge was converted to 2ft 9in and the line extended, involving passage under the Chain Pier, making it three quarters of a mile long, with a stopping place half way. A new car was provided with a 10 hp motor placed under the centre of the floor and the power supply was replaced by a more powerful 12 hp Crossley gas engine driving a Siemens dynamo located in a cave opposite the Eastern Terminus. The new track was supplied by Kerr Stuart & Co. of London and Glasgow. The rebuilt line was opened on 4th April 1884 but the first car was overloaded with dignitaries and needed to be replaced by a second car, later in 1884.

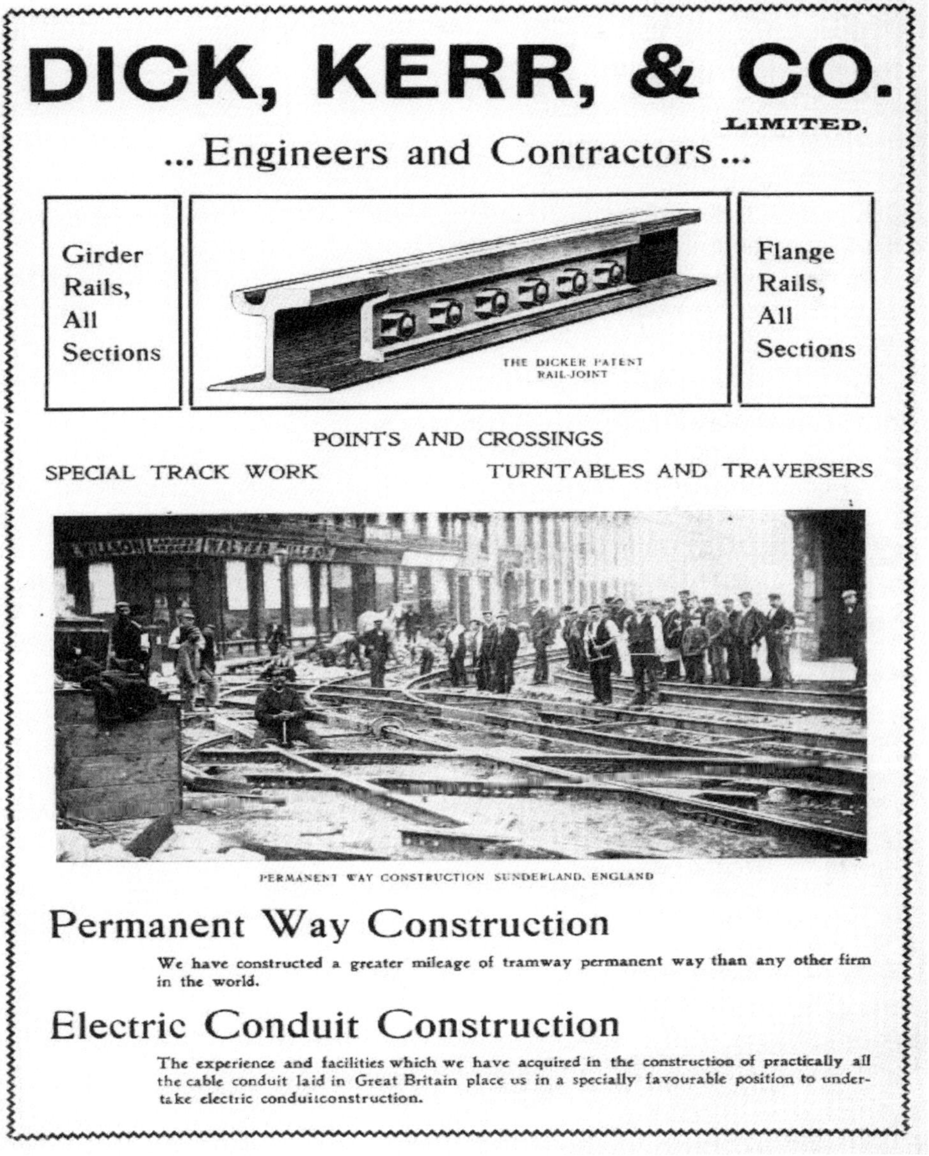

Dick, Kerr Advertisement from 1900.

Initially the current for the motors was collected from the two insulated running rails 'at 100 volts'. This was later modified to use current collected from a third rail with the return circuit through the running rails.

At a later date Magnus Volk sought advice from Karl Moritz Immisch over experiments to create an electric horseless carriage. A three wheeled prototype was built in 1888, powered by a 0.5hp Immisch motor.

At the end of 1888 and during 1889, the Electric Traction Company, employing Immisch machinery and expertise, instigated a trial of accumulator tramcars on the Barking Road section of the North Metropolitan Tramway Company's network. This small mile long single track from Plaistow to Canning Town was chosen to prove the economy and reliability of the electric system. The 6 tramcars ran daily from June 1889 until August 1892.

Giant's Causeway and Portrush Tramway:

The Giant's Causeway Portrush & Bush Valley was an eight-mile line to the 3 ft gauge, opened between 1883 and 1887. This used 245 V d.c. but at the start and until 1899 the current was collected from a side third rail. Public outcry and the general fear of widespread electrocution caused this third rail to be prohibited on the mile or so of street running through Portrush, and so steam tram locomotives had to be used on this length until the overhead wires went up; the steam trams remained in service to assist on peak days until 1915. On the opening day and in response to continuing public concern over the live rail the chairman of the company, determined to stop any further trouble, sat down on the live third rail before the assembled crowd and made a lengthy speech while in that position. Entirely reassured, the passengers climbed aboard the train for the inaugural journey to Bushmills, while the chairman, having waved them off, limped home to have some nasty burns seen to. The railway, complete with its original power station survived until 1949.

Dick, Kerr & Co. was responsible for the electrification of the line including the contact rail system, the supply of rolling stock and power supply equipment.

The Northfleet Conduit:

The first electric tramway in this country, in which the cars were run in series, was opened in March 1889 at Northfleet, where a line was constructed, to the 3ft 6in gauge, by the Series Electric Traction Syndicate as a continuation of an existing horse tramway.

In 1888 The Series Electric Traction Company of London were looking for somewhere to experiment with a new type of powered tramway to replace that of the horse drawn trams. They were aware of plans to extend the Gravesend to Northfleet line and acquired the powers granted in 1884 for the extension from the 'Leather Bottel' through Northfleet, to the gates of Huggins College. A contract was signed on 29th August 1888 for them to construct the line and operate it 'according to the Short-

Nesmith System' for which they held the rights. The spring jacks, switchgear, lighting and sundry apparatus, was supplied by the United Electrical Engineering Company. The whole of the road work and mechanical details were designed by Mr. Joseph Kincaid, M. Inst. C.E., the company's consulting engineer, and the construction was executed for him, under the personal supervision of Mr. J. E. Waller, A.M.I.C.E., by Messrs. Dick, Kerr, and Co. All the electrical work was carried out under the supervision of the company's electrical engineer, Mr. E. Manville.

The Short-Nesmith system used a positive and negative cable laid in an 8inch diameter conduit which was connected to the car by bars which passed through a slot immediately inside one of the running rails. At intervals of 21ft the supply cable was broken by a pair of plates held together by springs which were parted by the collector as it passed along the conduit, thus connecting the motor of the car to the supply and return cables; the collector was always in contact with one pair of plates. The current flowed in series from one contact to the next, through one side, through the motor of the car, and back through the other side. By this means any number of cars could run on the line at any one time. The electrical supply was generated by a 30,000 watt dynamo supplied by J. G. Statter & Co., with only one car running the current consumption was 60 Amps at 165 Volts.

The line was inspected by the Board of Trade on 26th March 1889 and the tramway was in operation from 16th April 1889. The experiment lasted for about 18 months and by November 1890 the electric trams had ceased operation and the line was converted for use by horse cars. The company having pulled out turned their attention to different experiments elsewhere.

The Gravesend and Northfleet Electric Tramways Ltd. was registered as a limited company on 18th April 1901. The system was reconstructed and extended for electric traction and in 1902 the first 20 cars for the new tramway were supplied by The Electric Railway & Tramway Carriage Works Ltd., Preston. Electrical equipment for the cars was supplied by Dick, Kerr & Co. Ltd.

Dover Corporation Tramway:

By the 1890s the town of Dover had grown considerably to the north and west and some way to the east below, and northward from Dover Castle; but no attempt had been made to provide passenger transport, except by privately-owned horse omnibuses. On 9th November 1895 at a special meeting of the Dover Town Council a resolution was proposed by Councillor G. F. Fry and seconded by Councillor Barnes – that application be made to the Board of Trade for a provisional order authorising the construction by the Corporation as the local authority of tramways in the borough, of the 21 members present at the meeting, 19 voted in favour. At the time Councillor Fry was Chairman of the London Street Tramways Company and in 1898 became a director of the newly founded Electric Railway and Tramway Carriage Works in Preston.

By July 1896 a provisional order had been made and confirmed by the Board of Trade and various modes of traction were under consideration. Gas-driven trams were fully investigated but it was decided to adopt electric traction, as recommended by the consulting engineer, Mr. Stephen Sellon, A.M.I.C.E. Tenders were invited and that of Dick, Kerr and Co. Ltd. was accepted for the supply of electrical equipment and rails and that of J. J. Briggs & Co. of Blackburn for the construction of the permanent way. On 6th September 1897 Dover became the first town in the south of England to introduce electric trams. A gauge of 3 ft 6 in had been agreed, together with an overhead-wire system, and when completed covered a distance of 3 miles.

In 1897 Dick, Kerr and Co. Ltd. had not yet entered into the business of building trams, it would be another 18 months before they would be able to supply trams made by the Electric Railway & Tramway Carriage Works, in Preston. Dover's first trams were supplied by Dick, Kerr and Co., Ltd., but the bodies, which were mounted on Peckham trucks, were built by the Brush Electrical Engineering Co. Ltd. at Loughborough. In 1902 and again in 1905 Dick, Kerr supplied cars, mounted on Brill Trucks, manufactured by the Electric Railway & Tramway Carriage works in Preston.

Blackpool & Fleetwood Tramroad:

In 1895 a syndicate was formed to promote a tramway connecting Blackpool with Fleetwood. Fleetwood was enthusiastic about the line but Blackpool Council preferred to keep complete control of the town's public transport after their take-over of the Blackpool Electric Tramway Company in 1892.

On 29th April 1896 it was reported that –

> It is highly probable that the Blackpool Improvement Bill will sail safely through Parliament with its main clauses intact. The Lytham and St. Annes Tramway Company and the Lancashire and General Tramways Company had agreed to withdraw their opposition on certain terms. The most important clause in the new Bill was that which asked for the absolute control of the municipal tramways to avoid any future need to advertise for tenders to lease them. The Company would be allowed to run their trams by any system as far as the South Shore Station only, paying at the rate of 8 per cent on the cost of construction. The Company would have running powers over the Blackpool line up to the Gynn, and would pay 10 per cent rental on the cost of the construction of that part of the line between the Gynn and the Borough boundary, near Uncle Tom's.

In May 1896 the Corporation agreed to grant a 21 year lease of the line to the Fleetwood Syndicate, and the Blackpool and Fleetwood Tramroad Act 1896 was then passed unopposed.

The prospectus of the company was published in early June 1897 –

The share capital was to be £120,000, with authorised borrowing powers of £40,000, and the prospectus stated --- that contracts for the permanent way and buildings had been let to Dick, Kerr and Co. Ltd. of Kilmarnock, and for the electrical equipment to Mather & Platt Ltd. of Salford Iron Works, Manchester. A contract was entered into with the General Construction and Tram-roads Company Limited, for these works, for the payment of the purchase moneys of the land, and of all preliminary and incidental expenses up to and including the allotment of shares, together with all professional charges, for the sum of £156,500. Under the provisions of their 1896 Act, the company may hold their line in perpetuity with the exception of the tramway in Fleetwood which may be purchased in 30 years by the UDC, and the tramway in Blackpool, which is held on a 21 year lease from the Corporation of Blackpool.

The work of construction which commenced in July 1897 was arranged in two contracts; Messrs Dick, Kerr & Co. Ltd. was to be responsible for the construction of the permanent way and buildings, and Messrs Mather & Platt were entrusted with the complete equipment of the power station, overhead work and rolling stock. The gauge of the track was to be 4 ft 8½ in.

George Flett was to be the resident engineer for Dick, Kerr and Co. Ltd., and Dr Edward Hopkinson for Mather and Platt Ltd.

Construction was placed in the hands of Benjamin Sykes and the financial arrangements were managed by Thomas Strover Turnbull chairman of the General Construction and Tramroad Company, with offices at 13, Spring Gardens, Manchester.

The General Construction and Tramroad Company Limited was wound up in January 1899, quote

At an Extraordinary General Meeting of the above named Company duly convened and held at the offices of Messrs. Sutton, Elliott, Turnbull and Mayne 13 Spring Gardens, Manchester on Wednesday the 21st day of December 1898 the following Special Resolution was duly passed and at a subsequent Extraordinary General Meeting of the said Company also duly convened and held at the same place on Friday the 6th day of January 1899 the following Special Resolution was duly confirmed –

That the Company be wound up voluntarily and that Mr. William Bolton 13 Spring-gardens Manchester Chartered Accountant be and he is hereby appointed Liquidator for the purpose of such winding up', signed T. S. Turnbull, Chairman.

The winding up of this Company coming so soon after the opening of the Blackpool Tramroad may be an indication that the reason for its existence was to finance the building of the Tramroad and almost simultaneously the founding of the Electric Railway & Tramway Carriage Works, which was established in Preston in April 1898.

The entrepreneurs who channelled their capital through this Company had a strong representation on the Board of Directors of the Blackpool Tramroad and of the Electric Railway & Tramway Carriage Works.

Turnbull raised the capital with no trouble at all and assembled a team of directors who between them already had interests in other tramway companies.

The Directors of the Blackpool & Fleetwood Tramroad Company as listed in the prospectus were –

Alexander Bruce, Esq., J.P., chairman Isle of Man Tramways and Electric Power Company, Limited.

Robert B. Barningham, Esq., 41, Victoria Buildings, Manchester.

Richard H. Prestwich, Esq., Holmleigh, Park Lane, Higher Broughton, Manchester,

George Readman, Esq. (director of the Scottish Life Assurance Company, Limited), Edinburgh.

George Richardson, Esq. (chairman the North Metropolitan Tramways Company and the Belfast Street Tramways Company), London.

The Secretary and Offices (pro-tem.) was Frederick Crimes at 13, Spring Gardens, Manchester.

Messrs Richardson and Barningham would also become directors of the City of Carlisle Electric Tramways Co. and Barningham with Turnbull was associated with the Manchester Traction Co. Ltd. with registered office at 13, Spring Gardens, Manchester, the office of T. S. Turnbull. Benjamin Sykes was also instrumental, along with John Kerr and George Flett, in establishing the Electric Railway & Tramway Carriage Works.

> Construction of the line commenced at the Fleetwood end on 19th July 1897, and there were some 400 navvies engaged on the work. A railway link was provided to enable a small locomotive to be used for transporting equipment and materials forward as the line progressed towards Blackpool. The most demanding sections of the line were between Anchorsholme and Little Bispham, where there was a deep cutting, and similarly at Tom's Cabin where there were cuttings up to 16 ft. deep.

At a January meeting of the Directors in Manchester, the Chairman George Richardson said that "It was hoped to open the line in May".

On 18th February 1898 the Fleetwood Chronicle reported progress on the construction of the line and supporting electrical distribution facilities.

At a half yearly meeting of the Company it was announced that –

> £82,000 had been expended up to January 17th the only works which were not progressing were those the Corporation has in hand.

John Cameron was appointed General Manager and Secretary of the company, and took up his duties in March 1898. He had been General Manager and Secretary of the Manx Northern Railway since 1879, and had acted as a consultant for the Douglas-Laxey electric line and, since 1895, to the Blackpool and Fleetwood Syndicate. At the first meeting of shareholders, as reported by The Isle of Man Times on 3rd September 1898 –

> Mr John Cameron, who as the late manager and secretary of the Manx Northern Railway Company, is well known and highly respected in this island, seems to be making himself as popular with his neighbours and invaluable to his employers as he was when at Ramsey. He is now, and has been for about six months the manager and secretary of the Blackpool and Fleetwood Tramroad Company, Limited. The shareholders of this company had their first meeting the other day, when it was reported that the business done was very good, over £6,000 having been earned in one month. Speaking of the manager, the chairman of the company (Mr George Richardson, of London), said that "in Mr Cameron they had one of the best managers they could possibly have. He was one of the most indefatigable men he knew".

The first passenger journey over the whole system took place on 1st July 1898 when crossbench car 4 travelled up to Talbot Road Station to collect a distinguished party of Blackpool councillors, officers of the Tramways Department and the press. The Board of Trade inspection took place on the morning of Wednesday 13th July. Two cars were provided at the Blackpool terminus (Dickson Road) for the 'company of gentlemen' who were to participate in the trial trip. Major Marindin and Major Cardew travelled by train from London, the former was to inspect the Tramway generally, and the latter the electrical aspects; Major Marindin C.M.G., R.E., was the Senior Inspector for the Board of Trade –

> The cars had travelled no further than Springfield Road when the attention of the Inspector was drawn to the levels of the road and when the incline to the top of Dickson Road was reached, and in front of the Park Hotel, a bus and railway parcels van were drawn up on either side of the road preventing the tram from proceeding further. Then at Warley Road two busses met each other and were unable to pass until they agreed which should pass first. Near the Duke of Cambridge Hotel a furniture van obstructed the way and the driver refused to move until he had finished loading up. The driver asked the Inspector if "he expected them to shift the van every time the trams passed". The Inspector, being amused by the comicality of the question responded "But you cannot place a furniture van in front of an Act of Parliament". The Inspector, in commenting on the problems found: "You will either have to widen the road or interlace it as you have done in other places. You cannot have an entrance to your Borough so narrow that the busses cannot pass. If you widen the road it will be alright".

> Mr Sykes pointed out "that they might widen the road and the bus proprietors could widen their busses and then the difficulty would be just the same"; the Inspector left it for the Corporation to resolve. During the course of the discussion it was found that the tramcars were eight inches wider than was allowed by statute. The Town Clerk declared that the Corporation would increase the width of the road in places while the tram company would decrease the width of the cars to that prescribed by their Act. As the defects were all connected with the borough section which had been constructed by the Corporation, and leased to the Company, and as the route from the Gynn to Fleetwood satisfied all requirements, the Inspector passed the line for immediate use, with a reservation as to the Blackpool end. The Corporation removed the objections by cutting away the kerbs, and on Thursday Major Marindin made a re examination and passed the closed section, and gave consent to the opening of the complete line to the public.

The first accident involving a tramcar occurred at Cleveleys only two days following when a lady broke her right leg in two places whilst jumping from the moving car.

The directors of the Blackpool and Fleetwood Tramway Co. celebrated the opening of the full route on Saturday 30th July by the escorting of parties of guests over the line during the afternoon and the holding of a banquet at the Imperial Hydro Hotel in the evening.

The Blackpool Times for Wednesday 3rd August 1898 covered the banquet in some detail –

> In honour of the historic event a number of gentlemen were invited by Mr. T. S. Turnbull, of Spring Gardens, Manchester, on behalf of the directors, to travel over the line in reserved cars, to inspect the generating station and accumulator houses, and to banquet at the Imperial Hydro. --- In the evening the guests were received in the reception room at the Imperial, at seven o'clock, and about half an hour later adjourned to the coffee room, where Mr. G. Richardson, of London, was installed as president. He was supported on his right by Councillor Jas. Ward (Blackpool), Mr. A. Bennie (Manchester) and Mr. G. Readman (Edinburgh); and on his left by Councillor R. C. Ward (Fleetwood), Mr. L. J. Davies (London), and Mr. J. E. Prestwich (Farnworth). Mr. T. S. Turnbull, Mr. R. H. Prestwich (Manchester), Mr. Jas. Greenwood (Manchester) and Mr. R. B. Barningham (Manchester) officiated as vice-presidents.

The guest list also included – W. B. Dick (London), George Flett (London), C. T. Cayley (London), W. J. Kirk (Edinburgh), F. Crimes (Manchester), J. Lancaster (Blackpool), John Kerr (Edinburgh), J. Rutherford (London), E. A. Stanley (London), John Cameron (tramway manager, Blackpool), A. H. Mayne (Manchester) and B. Sykes (Preston).

Blackpool and Fleetwood Crossbench Car, Fleet number 123, built in 1910 by United Electric Car Company.

Blackpool "Toast rack" Car. Built by the United Electric Car Company in 1912.

> Later in the evening one of the Speakers in response to the question – "to whom were they indebted for the success already attained"? – His answer– "the financiers and the solicitors, notably their good friend Mr. Turnbull and his colleague Mr. Mayne. On the Board of Directors too, they had men of capacity and experience and in the executive department they had such men as Messrs. Dick, Kerr and Co. and Messrs. Mather and Platt. They had also had the benefit of the all seeing eye of Mr. Benjamin Sykes under whose control everything had gone well and from whose observation nothing escaped".

The Tramroad, which was in effect a light railway, with fixed stopping places was 8½ miles in length and was completed in the relatively short time of eleven months and from the day the line opened , on 13th July 1898, up to 30th June in the following year, almost a million and a half passengers were carried.

Blackpool Corporation completed the required work on the street section and a regular service over the whole route began on 29th September 1898.

By Easter 1899, the fleet totalled 27 cars, including 13 crossbench cars, 3 crossbench trailers and 11 saloon cars which had all been built by G. F. Milnes of Birkenhead. The remaining 14 cars were built by The Electric Railway & Tramway Carriage Works in Preston and post 1905 by its successor The United Electric Car Co.

Blackpool Tram number 113 was built in 1914 by The United Electric Car Company Ltd. This single deck car had 48 seats and was mounted on UEC M & G type bogies with 2 x 40 hp motors and had a British Thompson Houston BTH B18 controller.

Blackpool and Fleetwood Tramroad Advertisement. Illustrating a Crossbench Car.

The first of these cars, built by The Electric Railway & Tramway Carriage Works, Nos 28-34, and known to the staff as 'Yanks' because of their Brill bogies and American-type equipment, were delivered in 1899. The United Electric Car Co. built further cars for Blackpool, 3 cars in 1910 and 4 cars in 1914, then following the take-over by English Electric the Dick, Kerr works continued to be a major supplier with 10 pantograph cars in 1928, 45 rail-coaches in 1933/4, 12 open boat cars in 1934 and 27 double deckers also in 1934.

By 1913 the company had gained the status of a fully-fledged railway company, they had fought in the courts for the right to be recognised as a railway, and pay only a quarter rating to the Local Authorities concerned.

The Company's lease on the Corporation-owned part of the Tramroad would expire in 1919 and a further 3 miles fell within the Borough of Blackpool following its amalgamation with Bispham-with-Norbreck UDC in 1917. This was perhaps the opportune time for the Directors to retire from the business for in 1917 the United Electric Car Company had been taken over by Dick, Kerr and Co. Ltd., Thomas Strover Turnbull, who had originally assembled the team of Directors and organised the financing of the project, had died on the 5th October 1917 and each of them were at an age when a more relaxed lifestyle would have been attractive.

A meeting of the Directors was held at Strangeways, Manchester (Boddingtons Brewery) on 5th March 1918, attended by J. Greenwood (Chair), R. H. Prestwich, R. B. Barningham, R. S. Boddington, all of whom had been Directors of the United Electric Car Company Ltd., and the Secretary John Cameron. The following was resolved –

That the undertaking be offered to the L & Y Railway Co. for £225,000, the Company to retain its outside assets and Cash and to pay its obligations, subject to compensation to the Directors and Staff for loss of Office.

The response from Blackpool Corporation was an improved offer of £284,000 including £4,000 for the Directors and £10,000 for John Cameron, backdated to 1st Jan. 1918. The takeover was enabled by the Blackpool Improvements Act of 1919, and in the interim the Company continued to operate the line.

Madrid and Barcelona:

In the summer of 1898, in spite of the war being waged by Spain, the streets of Barcelona and Madrid, the two most important cities of the Iberian Peninsula, 'presented scenes of great activity on account of the introduction of electric tramways in both cities by British capital'. Although the sympathies of Great Britain had been with the United States the engineers who were engaged in the work of tramway installation in Spain were not 'molested in any way, although most of them were Englishmen and among them were a number of Americans'.

The Tranvia de Madrid was owned by an English company, the Tramways Union Company, of London, who also owned and operated several other continental tramways. The company's lines in Madrid covered the principal streets of the city, and were approximately 17 miles in length. The Barcelona Tramways Company, Limited was also an English company, with offices at the same address in London. When the electric equipment of both lines was decided upon, Alfred Dickinson & Co. an English firm of Electrical Engineers, were appointed engineers to supervise the installation and tenders were requested from some thirty firms for the electrical equipment of both lines. In October 1897, the contract was finally awarded to Dick, Kerr & Co. Limited, of London, for the complete equipment required for the conversion of both systems.

On 16th November 1900, United States Consul-General Lay, writing from Barcelona reports on progress being made on the electric tramways in Barcelona:

1. Compania Anonima; owned entirely in England; completed and are running through the principal streets of Barcelona a double track of about 25 miles. The chief contractors were Dick, Kerr and Company, of England. The trolley wire, trucks, electrical machinery, engines and motors were all sourced from the United States.

2. Ensanche line (not named). The line was running horse cars and was owned by another English company, although the directors were those of the Compania Anonima, in conjunction with which it operated and was connected. The plans and specifications were in preparation, for the change from horse traction to electric, and it was expected that Dick, Kerr & Company would be its chief contractor.

A newspaper report of Thursday 18th July 1901 has a follow-up analysis of these contracts under the heading of Dick, Kerr, and Co. –

> Three years ago a British company secured a contract to supply the whole equipment of electric tramways in Barcelona and Madrid. No firm in this country could manufacture the articles required. The company executed the contract, but simply in the capacity of selling agents for American manufacturers. The firm which had this humiliating experience was Messrs. Dick, Kerr and Co., of Preston, who for a quarter of a century had been the largest tramway contractors in the country, and was well-known abroad, but their tramway work had either been horse or cable. Now, when a revolution in methods of traction had begun England was left behind. Our electrical engineering firms had gone to sleep. They did not foresee the growing new demand, or realise the tremendous possibilities of electric traction. To their credit, Messrs. Dick, Kerr, and Company determined to wipe out the disgrace which had fallen upon English enterprise proceeded to give an exhibition of 'hustling' which was stimulating. Two great factories at Preston rose into being at their bidding with mushroom-like rapidity. What was a bare field of fifteen acres in the June of one year was the site of the

best equipped electrical engineering works and car works in Europe in the following June. The firm which three years ago, was reduced to the status of a mere commission agent when dealing with electrical tramways depending on American manufacturers for its supplies can now supply 4,000 electric motors a year from its works and turn out four finished tramcars a day. It can take in hand the whole tramway undertaking of a town providing a complete service from beginning to finish, track, rolling stock, electrical equipment, generating station, and execute all the work within twelve months.

City of Carlisle Tramways:

There had been a proposal for a horse-worked tramway in Carlisle as early as 1880 but this never came to fruition.

Under the Carlisle Tramways Order of 1898 a contract was entered into by the Company with the Manchester Traction Company, Limited, who were the promoters of the Company, for the management of the Carlisle Tramways Order, and the construction of the permanent way, electrical equipment, and associated works and buildings, except generating station and plant. The contract for the construction work was awarded to Messrs Dick, Kerr, and Company, Limited, of London, and Kilmarnock. The following agreements were entered into:-

- An agreement dated the 16th June 1899, between the Mayor, Aldermen and Citizens of the City of Carlisle of the one part, and William Bass Needham, Enoch Horton, Joseph James Gittings, and Walter James Kershaw of the other part.
- An agreement dated the 22nd June 1899, between William Bass Needham, Enoch Horton, Joseph James Gittings, and Walter James Kershaw of the first part, the Manchester Traction Company, Limited, of the second part, and City of Carlisle Electric Tramways Company, Limited, of the third part.
- An agreement dated the 22nd June 1899, and made between the Manchester Traction Company, Limited, of the one part, and Dick, Kerr and Company, Limited, of the other part.
- An agreement contained in a letter dated the 22nd June 1899 from Dick, Kerr, and Company, Limited, to the Manchester Traction Company, Limited.

The Trustees for the Debenture Stock holders were - James Arthur Hutton, Esq., Alderley Edge and John Edward Prestwich, Esq., Farnworth, near Bolton.

The Directors were - Robert Breese Barningham, Esq. (Director, Blackpool and Fleetwood Tramroad Company), Manchester; Thomas Dewhurst Lingard, Esq., Burnside Windermere; George Readman, Esq. (Director, Blackpool and Fleetwood Tramroad Company), Edinburgh and George Richardson, Esq., London (Chairman, Blackpool and Fleetwood Tramroad Company and a director of the Provincial Tramways Group).

The Engineers were – Messrs Pritchard and Co. for the permanent way, and Messrs Alfred Dickinson and Co. for the electrical equipment, both of Birmingham.

The Company Solicitors were Messrs Sutton, Elliott, Turnbull and Mayne of 13, Spring Gardens, Manchester, also the address of Frederick Crimes, a joint Broker for the Company.

The Registered office of the Manchester Traction Company, Limited was also 13, Spring Gardens, Manchester; Messrs Barningham and Turnbull were directors of that Company and were also associated with the Electric Railway and Tramway Carriage Works and Barningham would become chairman of the United Electric Car Company. John Edward Prestwich would become a Trustee for the Electric Railway and Tramway Carriage Works and George Richardson its first chairman. Thomas Dewhurst Lingard would become a director of the English Electric Manufacturing Company.

Perhaps unsurprisingly one particularly interesting and revealing provision of the Carlisle Electric Tramways Company contract was that the 'Articles of Association of the Company provide that no objection shall be taken to any agreements entered into between the Manchester Traction Company, Limited, or any director thereof, on the ground that any of the parties thereto are promoters of the Company, or otherwise in a fiduciary position towards the Company, nor by reason of any of the directors of the Company being also a director of the Manchester Traction Company, Limited'.

The Manchester Traction Company Ltd., promoters for the tramway, was first registered on the 9th December, 1897, with a capital of £100,000 and an office at 13 Spring Gardens, Manchester, the office of Thomas Strover Turnbull, Solicitor; R. B. Barningham was also a director. Under the Carlisle Tramways Order, 1898 the promoters obtained powers to build and operate electric tramways within the city. The usual clause was included in the order, whereby under the tramways act, 1870, the Corporation may acquire the tramway undertaking at the end of 21 years, and thereafter at the end of every seven years, should they so desire.

The tramway promoters and their backers formed a new Company, The City of Carlisle Electric Tramways Company, Limited, which was first registered on the 16th June, 1899, and acquired the powers and rights of the 1898 Order. The registered office of the Company was at 23 London Road, Carlisle, and Mr. George Cook, was Secretary.

The joint engineers to the Company, Messrs. Alfred Dickinson & Company, and Messrs. Pritchard & Company, both of Birmingham, placed the whole contract for the construction of the tramways, the supply of track-work, overhead feeders and rolling stock with Messrs. Dick, Kerr & Company, Limited.

Work on the construction of the tramway started in Lowther Street on the 4th September 1899, by nine o'clock that morning the sets had been ripped up and a trench as broad as a cart-way had been made from Devonshire Street to Bank Street. So fast did the work proceed that the whole tramway was ready for inspection by the Board of Trade in June 1900.

The tracks were laid to 3 foot 6 inches gauge, 83 lbs. to the yard rails being used, which were 6 inches in height and rolled in Germany. Steel fishplates were used weighing 50 lbs per pair and embodying the patent 'Dicker' spliced rail joint. Messrs. Dick, Kerr laid out all the special track-work and junctions at their Kilmarnock works before transport to Carlisle.

The Carlisle Tramways were inspected on the 28th June, 1900 by Mr. A. P. Trotter of the Board of Trade. The Inspecting Officers, all Officers of the Royal Engineers, would traverse and inspect the whole line and its equipment, prescribe speed limits and operating regulations they may think desirable in the interests of public safety. The system opened to the public on 30th June 1900, with a total of twelve tramcars, three single-deck and nine open-top double-deckers.

One of the double-deckers was used by Mr. Trotter for his inspection; amongst those with him in the car were Mr. T. D. Lingard, Chairman; Messrs. G. Richardson and R. B. Barningham, directors; Mr. Flett of Messrs. Dick, Kerr & Co.

The tramway was not a financial success; the shareholders got their last dividend of 1¼ per cent in 1906. Messrs. Dick, Kerr and company owned a share of the Company's capital and there were also associations with other syndicates, including the Blackpool & Fleetwood Tramroad Company, mainly through having common directorships.

Camborne and Redruth Tramways:

The initial proposal for an electric tramway to serve the Camborne and Redruth area came in 1898 to serve the various tin mines scattered throughout the area. An application was submitted by the Urban Electric Supply Co. Ltd. for a Light Railway Order for a tramway 8¾ miles in length between Camborne, Redruth and Portreath on the coast. This was rejected on opposition from the Great Western Railway which considered the tramway would be interfering in its area. The proposal materialised again in 1900 when a Tramways Order was granted for a system but only along the main road between Camborne and Redruth covering a distance of just three miles.

The tramway was constructed by Dick, Kerr & Co. in around 6 months. Track laying started on 7th April 1902 and was single tracked of 3 ft 6 in gauge with eight passing loops and double tracks at each end. The sharpest curve was 40 ft. radius and the steepest section 1 in 15. Lengths of rail and poles were landed at Portreath and then delivered to the various roadsides. The tramway was mostly complete by September 1902. Members of the Camborne and Redruth Councils visited on 1st October, and the Board of Trade Inspector passed the tramway for use on 25th October. The formal opening was held on 7th November and the tramway was opened for passenger service the same day.

The tramway became unique in Britain because one year after opening, in November 1903, a mineral ore service opened operated by two four-wheeled electric locomotives and 12 ore wagons sharing the track for part of its distance with passenger traffic. Vehicles were obtained from G. F. Milnes & Co. in 1902 and 1903.

The service closed to passengers on 29th September 1927 and in August 1934 the freight service for tin ore was replaced by an aerial ropeway.

The Hastings Electric Tramways:

The proposal for an electric tramway system came in 1897 from the Hastings, Bexhill and District Light Railways (Electric) Company Ltd. Lines totalling 12½ miles in length with a gauge of 3 ft 6 in were proposed covering Bexhill to Hastings and local routes within Hastings. The original plan was cancelled due to considerable local objections but fresh plans submitted in November 1899 by the Provincial Tramways Company, of which George Richardson was a Director, were passed as the Hastings Tramways Act of 1900. Despite continuing strong opposition to a line along the seafront, a line was finally agreed by the Hastings (Tramway) Extension Act of 1903 subject to two conditions, firstly, that the work would not begin until the construction agreed by the 1900 Act had been completed and, secondly, that the seafront route must not be worked by an overhead wire system.

Dick, Kerr and Company were the contractors for the whole system, including power house, sub stations, overhead work and track construction. Initially some 400 men were employed on the work; the men were paid 6d per yard and earned up to 4/6d per day if they worked the full day from 7.30 am to 4.30 pm with a half-hour break for lunch. Construction of the lines began in December 1904; car sheds were built, roads widened and within 8 months eleven miles of track had been completed.

The first 3 tramcars were delivered in early July 1905, arriving by train from Preston and carried from the station on trucks pulled by traction engines. The tramway opened on 31st July 1905 with 22 cars carrying 18,663 passengers on the first day.

In October 1905 representatives of the Engineer journal were invited by Messrs. Dick, Kerr, and Co. to journey to Hastings to inspect the tramway installation which had 'recently been put to work there by them' for the Hastings and District Tramways Co. The tramways were described as being remarkable in two ways; first by reason of the 'heaviness' of many of the gradients, and, next, because of the sharpness of many of the curves.

On 2nd September 1905 a fatal accident was reported concerning the new tramway –

> A young electrical engineer, Sydney Dixon, aged 21, of Bolton, employed by Messrs. Dick, Kerr, and Co., owners of the Hastings tramways, was on Monday explaining to an understudy which wires he could safely touch, when he placed his hand on a live wire of tremendous voltage and was killed instantly. His left hand was enveloped in flame when the understudy dragged him away.

Maidstone Corporation Light Railways:

The first ideas for a tramway in Maidstone came in 1880 when a private company put forward proposals for a horse tramway system. Plans were deposited in 1881 but the Bill was withdrawn and nothing further was heard. In 1900 consideration was given to an electric tramway and in May 1902 a application was made to the Light Railway Commissioners and permission was granted. The contract to build the tramway was placed with Dick, Kerr & Co. Ltd. of Preston at a cost of £18,000. An overhead wire system was used and these were supported by poles from both sides of the road. Six open-top four-wheel cars were provided, manufactured by the Electric Railway & Tramway Carriage Works Ltd. of Preston. The cars carried 22 passengers inside and 26 on top and each was equipped with 2 Dick, Kerr 25B motors of 25 hp each.

Rothesay Tramways:

The horse tramway was introduced to Rothesay by a provisional order under the Tramways Act of 1870 and the Rothesay Tramways Order Act of 1880; the tramway opened on 1st June 1882.

In 1899 the tramway was sold to the British Electric Traction Company and by the end of April 1901 the British Electric Traction Company had obtained full control of the Tramway Company and instructions were issued for the preparation of plans and specifications for the reconstruction of the tramways and Electrification of the line. Reconstruction commenced on 3rd March 1902 and by April the construction of the permanent way, including its conversion from 4ft to 3ft 6in gauge, was reported to be 'proceeding satisfactorily, being undertaken by Dick, Kerr and Company Ltd'. The track to the new gauge of 3ft 6in was fully reconstructed by early June; traction poles had been erected with the requisite overhead equipment and some of the new cars had been delivered and assembled.

The generating station, also built by Dick, Kerr, was equipped with English Electric Manufacturing Company generator of 150 kW capacity and driven from a Browett-Lindley vertical compound engine. Formal inauguration of the line took place on Wednesday 13th August 1902 although Board of Trade inspection was delayed until 16th and 19th August. A celebratory luncheon was held on 11th August with over 100 invited guests from Rothesay, Greenock and Glasgow. Dick, Kerr and Company Ltd. was represented by Mr. McAlpine their engineer responsible for the construction of the tramway.

Rothesay Electric Tramway. The tramway was built by Dick, Kerr & Co. and Opened on 13th August 1902. Car No. 13, one of the first batch of 15 cars, Manufactured by The Electric Railway and Tramway Carriage Works, Preston in 1902.

The first fifteen tramcars for the Rothesay Tramway were built by the Electric Railway and Tramway Carriage Works in 1902; these comprised ten Open Crossbench cars seating 50 and five Crossbench Cars seating 46. In 1908 a small Demi-car seating 20 was acquired from the United Electric Car Company.

Gustavus McAlpine, a young engineer aged 25 –

Joined Dick, Kerr and Co. on his return from the United States where from 1898 to 1900 he was in the Electrical Engineering laboratory at Harvard, and later with the American General Electric Co. engaged on the design and development of electric railway equipment. His early education was modelled on 'Herbert Spencer's system', followed by four years at Durham College, of which he was an Honours Medalist in engineering.

During his time with Dick, Kerr's he took charge of the conversion of the Rothesay horse tramways to electric traction, which included the erection of the power station. He also undertook work on the Dublin and Howth Hill electric railway, and the installation of the Durban electric tramways.

On leaving the Dick, Kerr Company he joined the Colonial Office, going out to Mauritius for three years as Electrical Adviser to the Crown Agents for the Colonies followed by five years in Ceylon. On the outbreak of war in 1914, he returned home and served first as Captain, R.E, then as Lieutenant, R.N.V.R., attached to the R.N.A.S. on night flying duties in France. In 1916 returned to the Admiralty, where as Lieut.-Commander, R.N.V.R., he was in charge of a group of electrical engineer officers.

After the war he was with the R.A.F. with the rank of Major then Squadron Leader. In 1925 he transferred to the Civil Service as a Senior Technical Officer in the Air Ministry with responsibility for research and development of electrical equipment for aircraft and aerodrome lighting, until he transferred to the Aeronautical Inspection Directorate in 1928, where he served until his retirement in 1933.

He was a Member of the Institution of Electrical Engineers and a Fellow of the Royal Aeronautical Society; he died in 1950 aged 75.

Preston Corporation Tramways 1904 – 05:

It would be remiss not to make some reference to the Preston tramway system, being the home town for the tramcar manufacturing business of Dick, Kerr & Company. A full description of the tramway and its construction, with maps and diagrams, is contained in 'Preston Corporation Tramways 1904 – 05 Official Description' a copy of which is held in the Harris Library in Preston.

The contract for permanent way and overhead equipment was awarded to Messrs Dick, Kerr & Co. Ltd., with special work by the Lorain Company, for £80,592. Some members of the Tramways Committee objected to the manufacture of part of the material by an American company (Lorain); although the value of this portion of the contract was only for £650. After some discussion the objection was withdrawn.

> The lines were laid over a route length of 10 miles of street; of these 6.3 miles was double track and 3.7 miles of single track, equal in all to 16.3 miles of single track. The steepest gradient was 1 in 16 and the sharpest curve had a radius of 42 ft. The track was laid to the Standard Gauge of 4 ft. 8½ in, the rails being British Standard section No. 2 weighing 95 lbs. per yard, and No. 2c weighing 101 lbs. per yard on the curves and manufactured by the North Eastern Steel Company. The rail joints were of the well known 'Dicker' type, the fish plates being 32 inches long, drilled to take eight 1 in. bolts, and weigh 73 lbs. per pair. The Special Work, points and crossings were made by the Lorain Steel Company and by Hadfield's Steel Foundry Co. Ltd.
>
> The Feeder Cables were manufactured and laid by the St. Helen's Cable Company, and were of the non-metallic sheathed class, insulated with the company's well known 'Dialite' dielectric double taped and braided with heavy jute. The feeder system comprises 7 positive and 6 negative feeders, varying from 0,5 to 0,25 square inches sectional area, and are drawn into Sykes patent stoneware conduits manufactured by the Albion Clay Company.

The overhead equipment was supplied and erected by Messrs R. W. Blackwell & Co. Ltd., as Sub – Contractors to Messrs Dick, Kerr & Co. Ltd. 924 steel poles were used in the construction, which were of two types, viz. Heavy poles weighing 1,240 lbs., and light poles weighing 840 lbs. made by Messrs Stuart & Lloyd of Glasgow.

The Trolley Wire was of hard drawn copper, 00 gauge, with a breaking strain of 5,000 lbs., and a resistance of 0,235 ohms per 1,000 yards; the sections varied from ⅓ to ½ mile in length.

The whole of the plans and specifications for the Preston Electric Tramways, including the Power Station Buildings and Plant, were prepared under the direction of Walter H. Tittensor, M.M.E.A., as Resident Engineer, who went on to officiate as Engineer and Manager of the Tramway. The work of construction was superintended by James Marshall, Chief Assistant Engineer for the Construction, and J. E. Foster, Architectural Assistant and Clerk of Works.

The car-shed and generating station were constructed at Holmrook Road, off Deepdale Road. The shed was built to accommodate 30 cars, and had six roads with entry from the special branch in Holmrook Road.

The installation of the power-station comprised equipment to the total value of £16,026, of which the major items consisted of:

- 2 Lancashire Boilers (hand fed) by Yates & Thom, Blackburn £1,591
- Cooling Tower and Condensers by F. Storey £2,212
- 2 300kW generators producing 500/550 volts d.c. by Dick, Kerr, with associated engines – horizontal cross-compound, built by Cale Marchant & Morley of Bradford. £5,916
- Switchboard by Ferranti £1,216

A large storage battery was also installed to supply current on a temporary basis in the event of a generator breakdown. In addition to steam being supplied from the main boilers, additional steam could be taken, as required, from the Corporation's 'Refuse Destructor Plant' in nearby Argyll Road.

At a meeting of the Town Council on Thursday, October 29th 1903 the tender of Messrs. Dick, Kerr & Co. Ltd. was accepted for the first batch of 30 electric tramcars at £472 per car. This was considerably lower than any other tender.

The cars were manufactured by the Electric Railway & Tramway Carriage Works in Preston and designed to comply with the requirements of the Resident Engineers specification for steel under-frames to be constructed without struts, to obviate the difficulty of 'sagging platforms' a problem, apparently met with in the ordinary wood under-frame constructed with struts. These first cars were all open-top, double deck cars, un-canopied and without vestibule with direct stairs.

Preston Corporation Tram. Built in 1904 by the Electric Railway and Tramway Carriage Works. Then in 1907 fitted with a top cover by the United Electric Car Co.

The original tender was for 30 four-wheel cars, but this was changed in January 1904 to 26 four-wheel and 4 bogie cars. The twenty-six four-wheel cars were mounted on Brill 21E 6 ft. trucks fitted with Dick, Kerr DK 25A, 25hp motors and had Spencer slipper brakes. These were numbered 1-26 in the Corporation fleet and had 3-window saloons.

The four, larger, bogie cars were mounted on Brill "maximum-traction" 22E trucks, fitted with Dick, Kerr DK 34A, 35hp motors and had hand slipper brakes acting on the wheels, and mechanically operated track brakes. These were numbered 27-30 in the Corporation fleet and had 4-window saloons and "landing" staircases. All the cars were equipped with 2 DB1 form C Special controllers.

The first electric tramway route to be opened was that to Fulwood Barracks via North Road and the first car (car number 24) departed from Lancaster Road at 10 a.m. on 7th June 1904.

Preston Tram No. 33 at the Factory. Built in 1912 by the United Electric Car Co., Preston.

Preston Tram No. 10 Passing the Old Militia Barracks.

Trams in London

There were several early attempts to run motor-powered trams on the London tramlines: John Grantham first trialled an experimental 23 foot steam tramcar in London in 1873 but withdrew it after it performed poorly. From 1885, the North London Tramways Company operated 25 Merryweather and Dick, Kerr steam engines hauling long-wheelbase Falcon Engine & Car Works trailers. Although several towns and cities adopted steam trams, the problems associated with track weight, acceleration, noise and power held back their general acceptance in London.

London County Council came into being on 1st April 1889 and in response to a growing demand for improved public transport, in a city with an ever expanding population, decided to exercise the power of compulsory purchase over the old tramway system. This could be exercised after 21 years from the date of authorisation of their construction and since the first tramway was opened in 1870, that time was near. By the late 1890s it was clear that the future of tramways lay in electric traction. To advise it on the best system to adopt, the LCC engaged Prof. Alexander Kennedy, a leading electrical engineer. Electrification of the LCC Tramways was in part delayed by the Council's insistence upon being its own power supplier, and to that end it was building its own station at Greenwich, which would not be ready before the tramway was completed. Therefore, on 9th July 1901 the LCC resolved to go into partnership with the South London Electricity Supply Corporation, which was registered on 4th December 1896 to supply electricity to Lambeth; its supply started on 28th November 1898.

Westminster, Waterloo and Blackfriars Bridges to Tooting:

In December 1901 London County Council placed contracts for the electrification of their tramways from Westminster, Waterloo and Blackfriars Bridges to Tooting. This line would be the first of the County Council tramways to be converted from horse haulage to electric traction.

The first contract was awarded to J. G. White & Company of London, for track laying and road work over approximately 14 miles of track, all on the conduit system.

A few days later the contract for the complete generating plants for the tramway was awarded to Dick, Kerr & Company of London for the installation of equipment manufactured by the English Electric Manufacturing Company Ltd. of Preston –

> The contract was for three phase 1500kW machines which included electric plant for two generating stations, one for a direct current temporary station to be erected near Loughborough Junction, and a three phase plant required for the permanent generating station to be erected at Greenwich, together with the necessary sub-stations in connection with the electrification of the tramway Messrs. Dick, Kerr & Company was awarded contracts for both stations.

> For the direct current plant 212 bids were received and that of Dick, Kerr & Company, for £25,964, was accepted. For the three phase plant Dick, Kerr & Company were accepted from 160 bids and although not the lowest quotation they were awarded the contract on technical merit, their price being £46,425. It was also seen to be a distinct advantage to have all the equipment, for both stations, supplied by the same manufacturer from the point of view of convenience and economy in work, superintendence and supply of parts. Dick, Kerr & Company undertook to deliver the direct current plant in ten months and the three phase plant within eighteen months from the date of the order. Dick, Kerr & Company sub-let the contracts for the main engines and auxiliary engines to Bellis & Morcom, Ltd. of Birmingham.

On 24th April 1902 the LCC considered nine tenders for the supply of 100 double-decks, double-bogie tramcars. Prices varied from £65,880 to £83,500, and delivery times from 24 to 70 weeks –

On the advice of Prof. Kennedy different suppliers were considered for different components:

- Trucks – J. G. Brill Co, USA.
- Car Bodies – Messrs. Milnes and the Electric Railway & Carriage Works.
- Plough – Messrs. J. G. White & Co.

The two lowest tenders which allowed for these were:

- The British Thompson-Houston Co. - £70,330.
- Messrs. Dick, Kerr & Co. - £71,754.

Prof. Kennedy, however, felt that Dick, Kerr & Co's motors "are the most powerful and give better acceleration than the others". Moreover Messrs Dick, Kerr & Co. was prepared to deliver the whole of the 100 cars within 24 weeks instead of the 36 weeks specified by other firms.

The first batch of cars for the new conduit system was delivered by mid-April 1903, and a new order for 200 cars was placed. The double-deck cars cost £659 and the smaller, single-truck cars cost £529.

The new conduit tramway between Westminster and Tooting was opened by HRH the Prince of Wales (later King George V) on Friday 15th May 1903 –

> Guards of honour of the Queens Westminster Volunteers and Metropolitan Fire Brigade, with engines, saluted Their Royal Highnesses on arrival at a specially-erected marquee at Westminster, and one of the Council's bands played the National Anthem.
>
> Mr Benn, chairman of the Highways Committee, presented an address to the Prince outlining the history of the London tramways movement.

Mr Gerald Balfour also spoke of the importance of cheap and easy means of communication between town and suburbs, incidentally mentioning that he considered the conduit system of electric traction adopted by the council to be in the experimental stage.

The Prince of Wales then said – "Ladies and Gentlemen, at the request of Lord Monkswell, the chairman of the council, I have much pleasure in declaring the Westminster and Tooting section of London County Council electric tramway to be open for the use of the public".

The first tram was started by the Prince of Wales and was then driven to Tooting by the Tramways Manager, Alfred Baker. Mr John Williams Benn, the Chairman of the LCC Highways Committee, acted as conductor collecting the halfpenny fare and issuing special tickets printed in blue and gold. Also present on the tram were the Princess of Wales, her two small sons Edward and Albert,

The new conduit tramway between Westminster and Tooting was opened by the Prince of Wales on 15th May 1903. The first tram was started by the Prince and then driven to tooting by the Tramways Manager. The tram, number 86, was one of a batch of 100 manufactured by the Electric Railway and Tramway Carriage Works.

Mr. Gerald Balfour, President of the Board of Trade, and Lord Monkswell, Chairman of the LCC. At Tooting the party visited local workmen's housing built by the LCC and then they returned to Westminster on the top deck of the tram.

The vehicle used was number 86 of class A, being one of a batch of 100 ordered from Dick, Kerr of Preston. The bodies were built by the Electric Railway and Tramway Carriage Co. Ltd. They had Dick, Kerr DB1 form D controllers and a pair of Brill 22E bogies, each with a 30 horsepower Dick, Kerr 3A4 motor. One bogie in each pair was modified to take the plough for the conduit current collection used by the LCC. The cars seated 66, 28 inside and 38 on top. As built the trams were open topped, but within the first few years received top covers. For the opening of the tramway car number 86 was painted white and was decorated with evergreen fern. Inside, the wooden seats were replaced by pale blue and pink armchairs and, to complete the luxury, blue and white curtains and blue carpet had been fitted

The Kingsway Subway:

In 1898 land cleared around Holborn, next to the City of London, created an opportunity to use some of the new streets to be formed for tramways linking north and south London. At the same time it was also suggested that, if the trams were to run below the streets, in the manner recently adopted in New York, this would free the new thoroughfares for other traffic. A small deputation duly visited New York and Boston to see their subways at first hand. The party reported back on 5th November 1901, and on 12th November the Council resolved to seek powers for the construction of a subway for single-deck tramcars; The line was to commence at Theobald's Road, proceed along the level until Southampton Row was reached; then it would go below the level of the street in a subway varying in depth from 3ft below the surface to a little more, and pass along the site of the proposed new thoroughfare from Southampton Row to the Strand. The line would then turn towards the Embankment, which it would reach near Waterloo Bridge, where it would emerge into the open. After numerous political and technical hurdles had been overcome the line eventually came into use, initially for driver training on 30th December 1905.

The width of the subway was 20 ft. and the height 13 ft. 6 ins. so that a different type of car, to those with which Londoners were familiar, would be needed. The new tramcars were designed by Aubrey Fell, the Chief Officer of the tramway and built by the United Electric Car Co. Ltd., Preston. To comply with Board of Trade regulations for the construction of cars for tube railways, the trams had to be made entirely of non-flammable materials. They were mostly made from steel. The inside finish of the cars was composed entirely of aluminium, on which suitable designs were chased and picked out in colours.

Sixteen of these cars were ordered; known as Class F and numbered 552-567 in the LCC fleet. A later order for 34 cars was given to Brush Electrical.

London County Council Car, number 560, emerges from the Kingsway Tunnel. This was one of 16 Class F cars manufactured in Preston by the United Electric Car Company Ltd. They were mounted on 2 x M&G trucks and had Dick, Kerr motors and controllers. In order to meet Board of Trade regulations these cars had to be made of non-inflammable materials.

Tramways on the Thames Embankment and Over the River:

This extension was needed to connect the southern terminal of the subway adjacent to Waterloo Bridge with the tramways on the south bank of the Thames.

Reported on 13th May 1902 –

> A parliamentary Committee to-day decided that the construction of an electric tramway be authorised from Westminster Bridge to Waterloo Bridge.

On 15th May 1902 –

> The London County Council has got its tramway Bill and in the near future electric cars will be running on the Victoria Embankment. This has been a pet project with the authorities at Spring Gardens for many years, and on nine previous occasions they had unsuccessfully endeavoured to get Parliamentary sanction for the scheme. Their proposal has been characterised as vandalism, and many pathetic appeals were made to Parliament to 'protect the beautiful boulevard from the hand of the spoiler'. It was admitted that the tramway would be of great service and as the council had undertaken to run the cars on the underground (*conduit system*), instead of the overhead system, there was little likelihood of the beauty of the thoroughfare being spoiled. A suggestion had been made that the track of the tramway should be turfed. This plan had been adopted in the suburbs of Boston, with the result that the strip of well-grown grass in the centre 'considerably

> enhances the appearance of the streets, and deprives the tramways of much of the hideousness that may otherwise attach to them'. The Council considered this proposal, and though there were some difficulties in the way of its application to the Embankment, they were 'believed to be not insurmountable'.

The tramway continued to stimulate much debate and press coverage, including suggestions for alternative schemes, which delayed further progress. On Saturday 9th May 1903 one imaginative correspondent proposed –

> A Rail-less tramway involving enlarged motor-cars attached to a single wire from overhead wires sufficiently long for a car to thread its way among the ordinary traffic. No rails are required, and it is to be seen in France in daily use.

In 1903 a serious investigation into the state of London's traffic was considered necessary and a Royal Commission on the 'Means of Locomotion and Transport in London' was set up and published its findings in 1905.

On 4th April 1905 –

> Trams on the Thames Embankment – At the evening sitting the House settled down to consider the London County Council Tramways Bill. Mr. Whitmore moving an instruction in the committee dealing with the measure to omit those portions of it which proposed to sanction lines over Westminster and Blackfriars Bridges and along the Thames Embankment. The case put forward by the hon. Member and those who supported him was that it would be premature to sanction those lines before the Royal Commission appointed to consider the question had made its report.

On 27th May 1905 a Select Committee of the House of Commons approved the bill for the construction of tramways over Blackfriars and Westminster Bridges and along the Thames Embankment.

Reported on 2nd June 1906 –

> The London County Council had secured a victory as regards its bill for the construction of tramways along the Victoria Embankment and over Westminster Bridge and also over Blackfriars Bridge. The bill was ordered for a third reading. The House of Lords was not expected to offer such strenuous opposition to the bill as in past years, especially as the Corporation of London had agreed with it.
>
> The committee insisted that pending the widening of Blackfriars Bridge, the tramway on the Embankment would have to stop at a point opposite John Carpenter Street, and the roadway at Westminster Bridge would have to be widened by taking 2 ft. away from each of the two footways. Clauses giving authority to the Metropolitan Police to control the traffic over the two bridges on state occasions were also inserted, and the period for construction was reduced from seven years to five years.

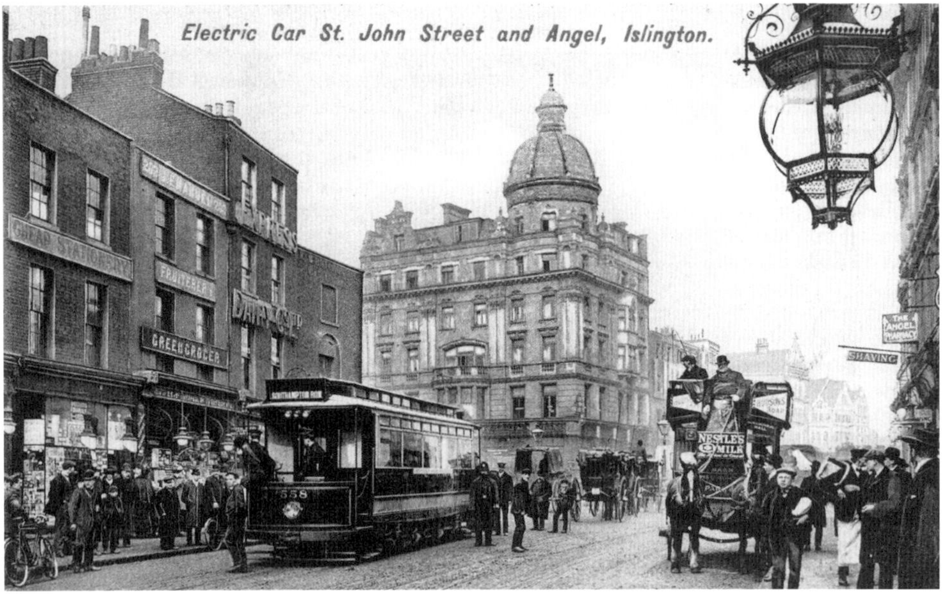

London County Council Class F Car 558. Built by the United Electric Car Company Ltd. in 1905. For Service through the Kingsway Tunnel.

Thames Embankment 1906. Electric Tramway, on the Conduit System for London County Council.

Trams running on the Conduit System over Westminster Bridge. Tracks laid by Dick, Kerr & Co. Ltd., and opened December 1906.

London County Council Class A Tramcars No's 68 and 75 built in 1903 by the Electric Railway and Tramway Carriage Works, Preston, travelling along New Cross Road.

> The Embankment and Bridges bill continued to meet with 'the most strenuous opposition', mainly from the bus and motor bus owners. It was, nevertheless, considered to be a bill that was 'bound to succeed in time, as it is one in favour of the general public, and there is a better chance now than heretofore with the strong Liberal Government now in office'. It had been revealed in committee that the tram lines on Westminster Bridge would not be placed in the middle of the bridge, but would keep to the side, also that the lines on the Embankment would be on the water side a few feet from the pavement. In this way they would interfere with carriage and bus traffic as little as possible.

By August 1906 Surveyors had been at work taking the levels of the roadway on the river side of the Victoria Embankment, in preparation for the work of constructing the electric tramways which were to link up the South London system by means of lines across Westminster and Blackfriars bridges. As late as September 1906 the project continues to excite public concern when the London County Council is accused of vandalism for intending to cut down the plane trees that line the Embankment. 'The trees are to be sacrificed to the trams'.

Reported in the Engineer for 12th October 1906 –

> London County Council was engaged, during 1906, on the work of laying down electric tramways on the Victoria Embankment. The tramways worked on the same conduit system as employed by the Council for its lines in other parts of the metropolis. The work presented considerable difficulty due, in part, to the hard nature of the road bed and to the fact that the vault of the underground railway tunnel came very close to the surface. However, the undertaking, the contractors for which were Messrs. Dick, Kerr and Co. Limited, made good progress.

Then on Tuesday 11th December 1906 and reported by the Gloucester Citizen –

> The London County Council may congratulate itself on the excellent progress that has been made with the laying of the Embankment tramway lines. The work involved a very heavy degree of labour at a period of the year when the weather rendered the conditions exceptionally unpropitious. Yet so rapidly was the task accomplished, that on Saturday next the line will be opened for traffic, and electric cars will be seen crossing Westminster Bridge for the first time. The junction of the Blackfriars and Westminster termini of the South London tramways will not, of course, be then completed owing to the necessity for widening Blackfriars Bridge, a work not yet begun. For the present the cars will only run up the Embankment to the north end of Blackfriars Bridge. It is anticipated that they will carry a great many passengers from the city to Westminster who now travel by the underground.

This report is typical, in that it gives all the credit to the authority, in this case the London County Council, and none to the constructor, the engineers, or workmen who actually accomplished this difficult task.

On 21st December 1906 The Engineer was able to report –

> The new line of tramways which the London County Council has laid along the Thames Embankment and over Westminster Bridge was opened to the public on Saturday last. Although it seems to work satisfactorily so far, much inconvenience is caused to other vehicular traffic at the foot of the bridge on the Lambeth side by the crossing of lines from the centre to the side of the roadway.

In June 1907 the Street Railway Journal reported –

> A very useful connection has just been made at the south end of Westminster Bridge, by which the Embankment tramways are now available for all of the cars operating on the route from St. Thomas Hospital, southward by way of Battersea and Wandsworth.

Finally on the 2nd October 1909 –

> The most important event of the year in the history of the London County Council Tramways is the formal opening of the widened portion of Blackfriars Bridge, over which the Council is operating its electric cars. This forms another and important link between north and south London, from a tramway point of view, and is the natural corollary to the opening, some time ago, of the Westminster Bridge to the tramways, and to the construction of the tramways along the Victoria Embankment. The formal opening of the bridge was attended by the Lord Mayor, Sir G. Wyatt Truscott, and Sheriffs, assisted by a large retinue of liveried servants. The Lord Mayor, assisted by Mr. Fell operated the first car across the bridge. The tramway lines and conduits had been constructed close to the western curb by Dick, Kerr & Company, as sub-contractors to Sir Wm. Arrol & Company, Glasgow who had widened the bridge at a cost of about £200,000.

Thus came to fruition a project first envisaged a decade earlier with the building of the Kingsway subway and the ambition to link the tramways on both sides of the river.

Metropolitan Electric Tramways:

An agreement between Middlesex County Council and the Metropolitan Tramways and Omnibus Company Ltd. (later renamed the Metropolitan Electric Tramways Ltd.) was signed on 16th November 1900 for the reconstruction and improvement of some 50 miles of main roads in the county. Metropolitan Electric Tramways Ltd. was registered on 15th January 1902 with four additional directors, one of whom was George Richardson, chairman of the North Metropolitan Tramways Company and since 1898 also chairman of the Electric Railway & Tramway Carriage Works in Preston. At this time North Metropolitan operated a large horse tramway, mainly over tracks owned by the London County Council and leased back to the company, but also operated over some lines in Middlesex. An agreement was reached whereby North Metropolitan would sell these lines in Middlesex to Metropolitan Electric Tramways, by exchange of shares.

The North Metropolitan introduced a Bill in the spring of 1902 to authorise the electrification of its own lines, the horse tramways from Finsbury Park and Stamford Hill to Wood Green and Edmonton; this received Royal Assent on 31st July 1902. Later in 1902 the Metropolitan Electric Tramways acquired a controlling interest in the North Metropolitan Company. Contracts were placed in November 1902, the main contractor being Dick, Kerr & Co. Ltd., who were to relay the track and carry out the necessary road widening for £92,788. Over the next few years Dick, Kerr & Co. Ltd, 'the acknowledged experts in tramway construction', would be awarded many contracts for the laying of tramways throughout the area covered by the Metropolitan Electric Tramways & Middlesex Light Railways, including – October 1904, Wood Green to Alexandra Palace and within Alexandra Park (£40,906) and from the horse tram terminus at the Royal Oak, Harlesden to Stonebridge Park (£36,338); May 1905, Highgate Archway to Archway Tavern (£4,588); May 1906, Great North Road Tramway Continuation (£21,652); July 1906, Wood Green to Winchmore Hill (£65,910) and Edmonton (£67,000); August 1906, Harrow Road & Paddington (£31,790);June 1908, Station Road, New Southgate to North Finchley (£31,437); July 1908, North Finchley to Childs Hill & Cricklewood (£116,978); July 1909, Ponders End to Enfield GER Station (£45,000) plus a number of smaller contracts amounting to about £48,000.

The 'G.B.' Surface-Contact Trial 1908/9:

In a report of 1899, Prof. Alexander Kennedy had looked favourably upon surface - contact systems of current collection, preferring them to conduit ones in that they were much cheaper to install.

Surface Contact System of Electrification being constructed.

He also noted that there was then "no large experience" of such systems in this country. Lincoln Tramways had operated the Griffiths-Bedell or 'G.B.' surface-contact system since 23rd November 1905 and the firm that made it - G.B. Surface-Contact Co. Ltd – were based in London. Early in 1908 the LCC decided to try the system for itself, and Dick, Kerr & Co converted the equivalent of 5.9 miles of single track in Mile End Road and Bow Road, between Bow and Aldgate.

On 21st June 1908 the Board of Trade inspected the system, and the following day gave it a six month certificate, permitting trial runs to begin. After a few stutters, which included the electrocution of a horse on a permanently live stud (*the horse recovered*), the trials were completed on 23rd March 1909. The LCC cancelled further trials converting the track to work partly on the conduit and partly on overhead wires, and reopened it on 31st July 1909.

Sir Alexander Blackie William Kennedy –

During his early years Sir Alexander Blackie William Kennedy, LL.D. held the appointment of Professor of Engineering at University College, London, taking up this post at the age of 27. He achieved distinction in this capacity by establishing the first college engineering laboratory. His practical ability brought him a great deal of consulting engineering work, so that in 1889 he resigned his professorship and occupied himself wholly with design and construction work.

He turned his attention to electrical engineering, and in a few years had built up one of the largest consulting engineering practices in the country. In 1894 he was elected President of the Institution of Mechanical Engineers and devoted his Presidential Address to the generation and transmission of electrical power, which was, by then, becoming his chief interest. At about this time he became chief engineer for the Westminster Electric Supply Corporation; he was also chief engineer from its commencement of the Central Electric Company and was engineer to the St. James's and Pall Mall Electric Supply Company.

Kennedy was associated with the London County Council Tramways, and the Waterloo and City tube railway, the electrified lines of the Great Western Railway, and the Southern Railway, and the formation of the London Power Company. During the Great War he served on many technical committees. He was president of the Machinery Designs Committee of the Admiralty until the designs for the "Dreadnought" were settled; in 1913 he was a member of Lord Parker's Committee on Wireless Telegraphy and shortly before his death, on 1st November 1928, was chairman of the Electric Railway Advisory Committee.

In 1906 he served as President of the Institution of Civil Engineers and was a Fellow of the Royal Society.

Electric Railway & Tramway Carriage Works, Strand Road c1898 In the foreground the site of the soon to be built West Works of the English Electric Manufacturing Co. Ltd.

A TRAM-WORKS IN PRESTON

By the end of the 19th century Dick, Kerr & Co. Ltd. had become firmly established and 'acknowledged experts', with a sound reputation, throughout the country, as constructors of tramways and light railways but the trams that ran on their lines were supplied by specialist tramcar manufacturers, such as G. F. Milnes & Co. and the Brush Electrical Engineering Co. Their work on the Blackpool & Fleetwood Tramroad (1896 – 1898) and the City of Carlisle Tramway (1899 – 1900) brought them together with the promoters of those projects and, perhaps most notably, Robert Breese Barningham and Thomas Strover Turnbull of the Manchester Traction Company. This company had been registered on 9th December 1897 with a registered office at 13 Spring Gardens, Manchester; also the address of Messrs Sutton, Elliott, Turnbull and Mayne, Solicitors. The company was wound up on the 30th July 1903 and removed from the register of companies in 1904, quote –

> At an Extraordinary General Meeting of the shareholders of the above named Company, duly convened, and held at the registered office of the Company, 13 Spring-gardens in the city of Manchester on Tuesday, the 14th day of July, 1903, the following Special Resolution was duly passed; and at a subsequent Extraordinary General Meeting of the shareholders of the said Company, duly convened, and held at the same place, on Thursday, the 30th day of July, 1903, the following resolution was duly confirmed, viz. :- 'That the Company be wound up voluntarily, and that Mr. W. Bolton, of 13, Spring-gardens, Manchester, Chartered Accountant, be and he is hereby appointed Liquidator for the purposes of such winding up'. - Sutton, Elliott, Turnbull and Mayne, 13, Spring-gardens, Manchester, Solicitors.

There is no evidence that the Manchester Traction Company was directly involved in tramway construction or tramcar manufacture. The most likely reason for their short lived existence (*and in the absence of documentary evidence this is speculative*) was to act as an investment broker for a small group of Manchester businessmen/entrepreneurs who were looking to invest in a new business venture, with reasonable expectations of showing a good profit. The participants in this enterprise were - Robert B. Barningham, Iron Merchant; Robert S. Boddington, Brewer and Richard H. Prestwich, Cotton Spinner who were joint promoters of the Blackpool and Fleetwood Tramroad Company, Britain's most profitable tramway. Barningham was also known to Dick, Kerr from his involvement with the City of Carlisle tramway and it is possible that it was these two ventures, and particularly Blackpool, which brought all the parties together to form the idea of a tramcar manufacturing business. Benjamin Sykes, as Supervising Engineer to the Blackpool & Fleetwood Tramroad Co., would have worked closely with Dick, Kerr & Co. on its construction.

Sykes, along with Henry Green, Frederick Crimes and Thomas Whitehead had, in 1871 on behalf of the North of England Railway Carriage & Iron Co., obtained from Preston Corporation the freehold on land occupied by the company on Strand Road, Preston. That company was wound up in 1878 and all equipment removed leaving the buildings empty.

All the interested parties needed to form a new business had therefore come together at Blackpool; Dick, Kerr & Co. Ltd., a tramway constructor wanting to expand into tramcar manufacture, a group of businessmen from Manchester with money to invest and a local businessman possessing access to land in an ideal location with a disused factory in need of development. The site had direct access to the railways and Preston Docks, for inward and outward transport of goods and, as a bonus was surrounded by unoccupied land for possible future expansion.

It is speculated that George Flett, on behalf of Dick, Kerr & Co. Ltd., along with Benjamin Sykes, as joint promoters of the new company, may have initially conceived the idea of entering into the business of tramcar manufacturing in Preston, although George Richardson, the first Chairman of the new company, may also have played a leading role. As a director of the Metropolitan Carriage & Wagon Company, Richardson had an established reputation for encouraging the British manufacture of tramcars. He was also Chairman of the North Metropolitan Tramway Company one of the first companies to order trams (horse drawn) from the fledgling tramcar manufacturer in Preston.

The Electric Railway and Tramway Carriage Works

The Electric Railway and Tramway Carriage Works Limited were registered as a company on 25th April 1898 as a subsidiary of Dick, Kerr & Co. of Glasgow.

From the Prospectus, published on 21st May 1898, the Company was incorporated with a share capital of £150,000 divided into 30,000 shares of £5 each. The initial issue of shares was for 22,000 shares of which the Vendors agreed to take 3,000 in part payment of the purchase money. The directors and 'their friends' subscribed for 6,000 shares, in the terms of the prospectus, 'which will be allotted to them in full'.

The Directors of the company were named as –

George Richardson, Esq., Chairman, North Metropolitan Tramways Company, Chairman.

George F. Fry, Esq., J.P., Director Brisbane Electric Tramways Company, Limited.

Richard H. Prestwich, Esq., Deputy-Chairman, Blackpool and Fleetwood Tramroad Company.

John Kerr, Esq., Director, Edinburgh and District Tramways Company, Limited.

George Flett, Esq., Director, Newcastle and Gosforth Tramways Company, Limited.

The company Solicitors were –

Messrs. Hope, Todd & Kirk, W.S. , 119 Princes Street, Edinburgh.
Messrs. Sutton, Elliott, Turnbull & Mayne, 13, Spring Gardens, Manchester.
Secretary (pro tem) - Mr. Frederick Crimes, 13, Spring Gardens, Manchester.
Registered Office - 13, Spring Gardens, Manchester.

John Kerr and George Flett represented the interests of the parent company, Dick, Kerr and Co. Ltd. of London and Kilmarnock.

The prospectus goes on to define the company objectives and details of the contract and agreements as follows –

> The Company has been formed for the purpose of manufacturing in the most complete and efficient manner all classes of cars and railway carriages, but especially tramway cars for horse, electric, and cable roads, and vehicles for light railways. The production of electric motor trucks and steel under-frames for Indian and colonial railways will also be a feature of the undertaking.
>
> The Company has also been formed to acquire valuable property and works in the West Strand Road, Preston, Lancashire, having an area of nearly 12½ acres, about 9½ acres of which will be used by the Company for the above purposes.
>
> The vendors have agreed to furnish and equip the works with machinery of the most modern character, and in particular, to lay down plant for driving the principal machinery by electric motors. They have also agreed to construct an overhead electric tramway system through the Works and grounds so that all the cars manufactured can be properly tested by the purchasers before leaving the works, thus enabling the cars to be put to work immediately on arrival at their destination. No other works in this country can offer similar facilities.
>
> It is expected that the works when completed will be capable of an output of 600 cars per annum at least, but the Directors are satisfied that even on a much smaller output the profits should be sufficient to pay satisfactory dividends. The companies with which the directors are actively associated have recently placed orders for over 300 tramway cars.
>
> The Company have taken powers to sell tramway cars and other rolling stock on the hire-purchase system. This class of business should prove a considerable source of profit to the Company.
>
> The Vendors, who are promoters of the Company, and are selling at a profit, have entered into a contract for the sale of the property and works, and the complete furnishing and equipment thereof, and for the payment of all the preliminary and incidental expenses and all professional charges up to and including the allotment of shares, and also all stamps, fees, law costs, and other expenses incident to the transfer and sale of the property to the Company for the sum of £80,000, payable

as follows:- £65,000 in cash, and £15,000 in fully paid shares, leaving available for working capital the sum of £30,000.

By the terms of the contract it is provided that the vendors shall hand over the works to the Company ready to commence operations as a going concern on or before the thirty-first day of October, 1898.

A valuation of the land and buildings, including the proposed alterations, was made by Mr. William Wilson, surveyor and valuer of Manchester and for the equipment of the proposed Works by Mr. Stephen Sellon, C.E., a well-known Electrical Engineer, both of whom submitted detailed reports. The Directors stipulated that the completion of the Works and the equipment of the machinery, in terms of the plans and specifications, should be carried out under the superintendence and to the satisfaction of Mr. Wilson and Mr. Sellon respectively.

The following Agreements were entered into –

(1) An agreement dated the 11th day of October, 1897, and made between Benjamin Sykes, Thomas Whitehead, and Henry Green, of the one part, and Frederick Crimes of the other part.

(2) An agreement dated the 24th February, 1898, and made between the Corporation of Preston of the one part, and the said Frederick Crimes of the other part.

(3) An agreement dated the 22nd April, 1898, and made between the said Frederick Crimes of the first part, George Flett, John Kerr, and Benjamin Sykes, of the second part, and Frederick William Browne on behalf of the Company of the third part.

Although not specifically named as such this, along with evidence from the initial share allocation, would imply that Messrs. Flett, Kerr and Sykes were the Promoters and Messrs. Flett and Kerr, on behalf of Dick, Kerr and Co. Ltd., along with Thomas Strover Turnbull, representing the Manchester Group, were the Vendors.

The Blackburn Standard for Saturday 25th March 1899 reports –

A new Industry for Preston:

Preston is to be congratulated upon having a new industry established in its midst. The formal inauguration of the Electric Railway and Tramway Carriage Works, Limited, took place on Thursday. The Company's new works cover three acres of ground, and are expected to turn out as many as 600 cars a year, operations having been begun with 200 workmen. Other Lancashire boroughs which would gladly see their local industries more diversified will be inclined to envy Preston its turn of fortune.

The Engineer journal for 31st March 1899 reported –

> On Thursday last a large new factory was opened at Preston, by the Electric Railway and Tramway Carriage Works, Limited. These new works, which occupy the site of the old Preston Carriage and Iron Company's premises – which have been disused for a number of years – are close to the Albert Edward Docks and the River Ribble, and in direct communication with the London and North-Western and the Lancashire and Yorkshire Railways, so that they possess every facility for the receiving and despatch of materials and goods. The old buildings have to a large extent been reconstructed, and with other shops that have been added, the buildings now cover four to five acres, whilst about thirteen acres altogether are enclosed by the works.

The report went on to describe the facilities and manufacturing processes in some detail, and then continued –

> The company, besides manufacturing trucks for electric cars, is also prepared to supply and fit to the car bodies the trucks of American manufacturers, such as Brill and Peckham. They are in a position to turn out two cars per day, or, on the basis of the ordinary size of tramway car, with roof seats, the capacity of the works is about 800 cars per annum. Large cars of the eight-wheel type, both for electric railways and tramways, are to be turned out when required, and the company will also be in a position, if necessity arises, to build railway carriages. The number of men employed will range from 600 to 800. In celebration of the opening of the works a number of visitors, together with deputations from the various municipal and local authorities throughout the country, were invited to a general inspection of the different shops, after which luncheon was served, the chairman of the company, Mr George Richardson, presiding; and the Mayor of Preston, in proposing the toast of the Electric Railway and Tramway Carriage Works, remarked that "they were undoubtedly the most important of their kind in the kingdom".

In March 1899 the company advertised for –

> Body Maker, Car Finishers, and Cabinet Makers; to good and steady reliable men steady employment and highest wages; working hours 50 per week – Apply personally or by letter, to the Electric Railway and Tramway Carriage Works Limited, Strand Road Preston.

In June 1900 and again in July 1902 the company exhibited, in conjunction with Dick, Kerr and Co. and the English Electric Manufacturing Co., at the Tramways and Light Railways Exhibition, held at the Royal Agricultural Hall, London.

Electric Railway and Tramway Carriage Works advertisement from 1899.

1900 Tramway and Light Railway Exhibition, Agricultural Hall, Islington, London. Joint exhibit Dick, Kerr & Co. Limited, English Electric Manufacturing Co. Limited and the Electric Railway and Tramway Carriage Works Limited.

Establishing the Business

The earliest available records of the company provide an interesting glimpse into the processes involved in establishing the Strand Road works as a viable business and the names of the senior people involved.

- On 30th July 1898 a transfer of shares is approved, G. Flett, 750 shares and John Kerr, 750 shares with the National Bank of Scotland.
- On 24th August 1898 it was agreed to apply for space at the Paris Exhibition.
- On 7th October 1898 a resolution is passed for the payment to Messrs Sykes, Kerr and Flett (*the promoters and vendors*) of £20,000 on account, contract as per Mr Williams certificate. Various cheques were approved by R. B. Barningham and R. S. Boddington, Directors.
- On 3rd November 1898 A. E. Stanley (*formerly of the Jackson & Sharp Company of Wilmington, Delaware*) was appointed Manager and John Baynes, Secretary.

Robert Breeze Barningham and Robert Slater Boddington were given the authority to sign letters to the banks.

The Company acquired an adjoining plot of land for £1,000.

- On 2nd December 1898 John Anger was appointed Foreman over the Wood Department. John Browell was appointed Foreman over the Fitting Shop. At this meeting it was also noted that working hours should be considered.
- On 9th May 1899, proposed arrangements for the opening of the Works were considered.
- On 27th July 1899, Preston Corporation had agreed to a reduction of rent from £476-17-2 to £340-0-0.
- On 15th February 1901 it was agreed to repay a loan of 10,000-0-0 from the English Electric Manufacturing Company.

During 1901 the accounts show regular monthly payments to J. G. Brill and Co. of amounts ranging from £ 689-6-8 to £6,000-0-0; this American manufacturer of Tramcar equipment was, at this time, a major supplier to the Company. From December 1903 there were regular monthly payments of significant amounts to Docker & Co. (Docker Brothers Ltd.), Gabriel & Co., G. H. Anger and Dick, Kerr and Co. Ltd., Kilmarnock.

Register of Members (The Companies Act 1862, Sec. 25)

The register of shareholders, in 1898, included the following –

Robert Slater Boddington	Strangeways Brewery, Manchester	Brewer	100 shares
Claud Thornton Cayley	Streatham	Engineer	200 shares
William Bruce Dick & Co. Ltd.	99 Cannon St., London	----------	100 shares
Edinburgh Investment Trust Ltd.	North Charlotte St., Edinburgh	----------	400 shares
Helen Margaret Kerr	C/O John Kerr, 94, George St., Edinburgh	Mrs	100 shares
Robert M. Kerr	119, Princess St., Edinburgh	Writer	30 shares
John Kerr	Luffness House, Drem, Haddingtonshire	Engineer	2000 shares
John Kerr	Luffness House, Drem, Haddingtonshire	Engineer	266 shares
Richard Henry Prestwich	Marsden Square, Manchester	----------	400 shares
George Richardson	98, Westbourne Terrace, Hyde Park	----------	100 shares
Walter Rutherford	110, Cannon St., London	----------	50 shares
Thomas Strover Turnbull	13 Spring Gardens, Manchester	Solicitor	100 shares
George Flett	110, Cannon St., London	Engineer	2000 shares

110 Cannon Street, London was the head office of Dick, Kerr and Company Ltd.

In the latter part of 1898 and in January 1899 the share transfer records show a considerable amount of share trading involving the principle shareholders including the Kerr family, George Flett and the family of Lewis E. Pickin with Sydney Howe Short amongst the transferees. The address given for John Kerr is variously given as 94, George St., Edinburgh; Luffness House, Drem or 110, Cannon St., London.

Debenture Trustees –

On 28th November 1901 an indenture is made (Companies Registration Office) between the Electric Railway & Tramway Carriage Works Ltd. of the first part, and John Edward Prestwich of Farnworth, and Thomas Strover Turnbull of 13 Spring Gardens, Manchester of the other part (hereinafter called 'the Trustees'). A mortgage also dated 28th November 1901 names the mortgagees and debenture trustees as John Edward Prestwich and Thomas Strover Turnbull.

The report from the Company for the year ended June 30th 1901 states that –

> After placing £4,000 to a depreciation fund, the profit for the year is £21,206, to which must be added £3,914 brought forward, making a total of £25,120. The directors propose to write off £7,342, for expenses incurred; to carry £2,000 to a general reserve fund; to pay a dividend of 8 per cent for the year; and to carry forward £3,778.

The Board sometimes held their meetings at Boddingtons Brewery, Strangeways, otherwise at the Cottage, Messrs Dick, Kerr and Co. Works, Preston.

A Report of the Directors to be submitted to the sixth Ordinary General Meeting of the shareholders to be held at the offices of the Company, Strand Road, Preston, on Thursday, the 22nd September, 1904, at 12 noon; the named Directors being -

George Richardson	Montagu Mansions, Portman Square, London, W., Chairman.
Richard H. Prestwich	Marsden Square, Manchester, Deputy Chairman.
George F. Fry	London Road, Dover.
John Kerr, M.P.	Gaddesden Place, Hemel Hempstead, Herts.
George Flett	Abchurch Yard, London, E.C.
Robert B. Barningham	Victoria Buildings, Manchester.
Robert S. Boddington	Strangeways, Manchester.

The directors submitted the balance sheet of the Company's trading for the year ending 30th June 1904 –

> After paying Debenture interest amounting to £2,500, and charging £4,209-7s-0d for depreciation , the profit for the year was £5,546-3s-4d to which was added £4,826-18s-7d brought from the previous year, making a total of £10,373-1s-11d.

The Directors recommended that this amount be dealt with, as follows:-

To pay a dividend of 5%, less income tax, for the year ending 30th June 1904 - £7,500-0-0; to carry forward to the next year - £2,873-1s-11d.

The Chairman went on to report that –

> "Severe competition and general slackness of trade accounts for the falling off in the profits for the past year; in the opinion of the Directors the prospects for the coming year have improved".

The sixth general meeting took place on the following day, 23rd September 1904, at which the Chairman in providing an explanation of the reduced dividend reported –

> The trade of the country had certainly not been so prosperous as heretofore, and especially in connection with this class of business. There had also been serious competition, but two of their most serious competitors had gone into liquidation, and the works had been sold, and consequently they were now out of the running, and they must look forward to getting better prices. He did not like to prophesy, but the directors thought they would not have an unsatisfactory balance sheet to present to them next year. One of the awkward things in connection with the business was the position taken up by the Federation of Engineers, which was, of course, the Trade Unions. They were now working with great zeal to do away with overtime and also to do away with bonuses for prompt and quick delivery. This latter practically meant that a bad workman must be in a position to earn the same money as a good one, and this was one of the points which would have to be fought later, particularly in all businesses connected with engineering.
>
> Messrs John Kerr, M.P., and Robert B. Barningham (Manchester), the two retiring directors, were unanimously re-elected.

The trading difficulties being experienced by the tramcar manufacturers at this time can be illustrated by the numbers of cars (approximate) being produced by the Electric Railway & Tramway Carriage Works, for the British market –

Total for the period 1899/1900	327
1901	597
1902	516
1903	421
1904	159

Despite short lived improvements, due partly to less competition, the market never fully recovers, the average yearly output for the period 1905 to 1913 amounting to 157 cars per year. Exports help to sustain the business but the works continue to operate below capacity until 1914 and the Great War.

AN AFTERNOON OFF was the headline on 29th May 1905 –

> A short strike at Preston Electric Works- The body-builders employed by Messrs. Dick, Kerr and Co. at the Electric Railway and Tramway Carriage Works, Strand Road, Preston, who number about 100, refused to return to work on Friday afternoon, in consequence of a threatened reduction in the price per car of about 10s. Satisfactory terms, however, were arrived at, and the men resumed work on Saturday morning on their former pay.

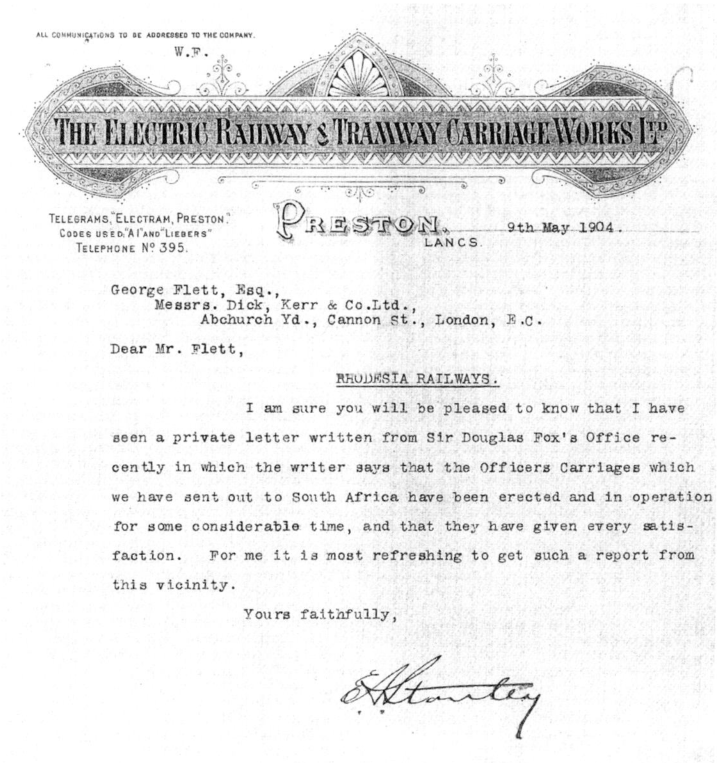

ALL COMMUNICATIONS TO BE ADDRESSED TO THE COMPANY.
W.F.

THE ELECTRIC RAILWAY & TRAMWAY CARRIAGE WORKS LTD

TELEGRAMS, "ELECTRAM, PRESTON."
CODES USED, "A I" AND "LIEBERS"
TELEPHONE No 395.

PRESTON, 9th May 1904.
LANCS.

George Flett, Esq.,
Messrs. Dick, Kerr & Co.Ltd.,
Abchurch Yd., Cannon St., London, E.C.

Dear Mr. Flett,

RHODESIA RAILWAYS.

I am sure you will be pleased to know that I have seen a private letter written from Sir Douglas Fox's Office recently in which the writer says that the Officers Carriages which we have sent out to South Africa have been erected and in operation for some considerable time, and that they have given every satisfaction. For me it is most refreshing to get such a report from this vicinity.

Yours faithfully,

Electric Railway and Tramway Carriage Works Limited Railway Carriage for Rhodesia.

Saloon Coach Rhodesia Railways. Built in 1905 by the Electric Railway and Tramway Carriage Works for the Administrator of Rhodesia.

Car for the Administrator of Rhodesia "Living Room".

Despite the occasional industrial dispute the company managed to earn a good reputation for building a quality product, for in The Engineer journal for 27th January 1905 there is a detailed report under the heading 'Saloon Coach, Rhodesian Railways' which describes in some detail, with illustrations the building of a saloon carriage for the use of the Administrator of Rhodesia, Sir W. H. Milton, for his journeying through Rhodesia. A letter from E. A. Stanley to George Flett dated 9th May 1904 also makes reference to 'Rhodesia Railways' – "I am sure you will be pleased to know that I have seen a private letter written from Sir Douglas Fox's Office recently in which the writer says that the Officer's Carriages which we have sent out to South Africa have been erected and in operation for some considerable time, and that they have given every satisfaction. For me it is most refreshing to get such a report from this vicinity".

George Richardson and the North Metropolitan Tramways Company

George Richardson, Chairman of the Electric Railway and Tramway Carriage Works, had gained a reputation as 'a father figure in the British Tramway Industry', being Chairman of the largest tramway company in the British Isles, the North Metropolitan Tramways Co., and a director of many other tramway companies including all those in the Provincial Tramways Group. He was aware of the high prices being charged by car-builders when demand exceeded supply, and became a director of the Metropolitan Carriage & Wagon Co., persuading them to take up tramcar building; in the meantime his companies bought from John Stephenson of New York.

The first tramway in London was opened in March 1861 by the American Entrepreneur George Francis Train, following on his earlier success in Birkenhead. This tramway ran along the Bayswater Road over a distance of just under one mile. Further lines were introduced, along Victoria Street in April 1861 then in August 1861 from Westminster Bridge to Kennington Gate. Further attempts to establish tramways in London between 1865 and 1869 all met with failure when successive Bills were rejected, although five tramways were operating in other parts of the country. Parliament anticipating powers being sought for more schemes passed the General Tramways Act of 1870.

In 1869 three new companies proposed tramways in London: north of the Thames there was the North Metropolitan Tramways Co., from Whitechapel to Bow; to the south were the Metropolitan Street Tramways Co., from Kennington to Brixton, and the Pimlico, Peckham and Greenwich Street Tramways Co., from Greenwich via Peckham and Camberwell to the south end of Vauxhall Bridge. The first to be running was the Metropolitan Street Tramways Co., on 2nd May 1870, followed one week later by the North Metropolitan Tramways Co. The passing of the Act of 1870 had opened the door, resulting in a surge in the number of tramway companies being formed, and tramways being constructed throughout the country. At this early date all tramways were horse drawn, and although by the late 1890s electric traction was becoming well established throughout the country, in London the horse continued to reign supreme, although other forms of traction had been tried without success.

Liverpool Corporation fleet No. 100. Open Top. Car built 1899/1900 by the Electric Railway and Tramway Carriage Works Ltd.

Liverpool Corporation fleet No. 208 on Lord Street. Built in 1900 by the Electric Railway and Tramway Carriage Works Ltd.

Electric Railway & Tramway Carriage Works, Open top cars for Liverpool Corporation Tramways, built in 1900, Transport by Road. The photograph taken on Strand Road, Preston. Near the entrance to Vine's Transport.

Electric Railway & Tramway Carriage Works, Transport by Rail.

Electric Railway and Tramway Carriage Works Blacksmiths Shop. Seat frames in right foreground.

In 1881 the introduction of the Faure accumulator rendered possible the application of stored electrical energy to the purposes of tramway work, and the first electric car made was run by Mr Radcliffe Ward on the Leytonstone line in 1882. M. Philippart, who had possession of the rights to the use of the Faure battery, conceived the idea of applying it at once to tramcars, and arrangements were made with Mr. George Richardson, then chairman of the North Metropolitan Tramways Company, to experiment on the Leytonstone line. M. Philippart had this, the first electric car, constructed in Belgium and sent over to England, and the working of the car was in the hands of Mr. Radcliffe Ward (ref: Scientific American Supplement March 2nd 1889).

In 1888 – 1889 a further experiment was carried out by the Electric Traction Company, employing the expertise of Karl Moritz Immisch, on the Barking Road section of the North Metropolitan network using the accumulator system.

The Electrical Engineer Journal for January 31st 1890 reported under the headline 'North Metropolitan Tramcars' –

> The following highly gratifying and encouraging remarks are from the half yearly report of the North Metropolitan Tramway Company. An agreement had been entered into with the Electric Traction Company for working the company's five accumulator cars on the Barking Road section, and the directors were able to report that the results during the half year had been satisfactory. The electric cars had run 34,366 miles and carried 469,095 passengers. The directors hoped that Parliament would sanction the introduction of electric power over the company's system, and enable them to extend its application to other lines. The original contract with the Electric Power and Traction Company had been for a fixed remuneration of 4½d per car mile run. When the contract came to be renewed the Electric Traction Company asked for an increase of their allowance to 5½d per car mile, this was granted but a short time later they asked for an advance to 7½d per car mile. The ensuing dispute over breach of contract which also involved the Electric Power and Storage Company, the makers of the accumulators, resulted in the tramway reverting to horse trams and the experiment with the accumulator ended.

In 1885 North Metropolitan experimented with steam traction but this was soon abandoned and the horse prevailed. In 1880 the North Metropolitan's horse stud had consisted of 2094 horses, by 1891 this had increased to 3355 horses and during that year 214 horses had died and 109 had been sold.

In 1892 and by the Act of 1871 the London County Council were authorised to purchase the tramway, the 21 year lease on the first section of track, authorised by Parliament, having expired. In 1893, at the 49th Ordinary General Meeting of the company, George Richardson was able to report that –

> They had carried in the year 82,627,342 passengers, showing the importance and popularity of the lines, also the Board would endeavour to put all possible obstacles in the way of the County Council acquiring the tramways at anything but a fair and reasonable price.

Opposition by the company to the takeover by the LCC continued into 1895, at the half yearly meeting of shareholders in January it was suggested by one shareholder that –

> "It was most iniquitous that the County Council should want to take over their property", "there was an election pending, and he hoped every shareholder in the company would have sense enough to protect his own interest by voting for those who would not take away their property".

Further opposition included North Metropolitan bringing an action for trespass against the Council, and in December 1895 taking the case to the Court of Appeal, the decision went against the company. On 24th June 1896 the LCC acquired the lines of the North Metropolitan Tramways Co. within the County, plus the whole of the London Street Tramways Co., but lacking the experience of operating tramways they were leased back to the company, who continued to operate them.

At the half yearly meeting of the North Metropolitan Tramways Co., in August 1903, George Richardson reported on the proposed electrification of the tramways although the "County Council had not yet decided on the system of electric traction to be used, but appeared to be in favour of the costly conduit system".

On 19th May 1905 it was reported that –

> The London County Council on Tuesday, by a large majority, decided to purchase the unexpired term of the lease of the North Metropolitan Tramway Company and its stock for £436,000, and to proceed with the electrification of the system. The ultimate cost of the acquisition of the London Tramway lines and their electrification was estimated at twelve million pounds.

The first order to be placed by the North Metropolitan Tramways Co., in 1899, with the newly established Electric Railway and Tramway Carriage Works, was, ironically, for 75 open-top horse drawn cars, this was not widely advertised by the Carriage Works since it had been founded primarily for the purpose of manufacturing electric cars, although the company's prospectus had allowed for the building of 'all classes of cars and railway carriages'. These were constructed in the original 1863 bays of the east works.

The Boer War

During 1899 - 1900 the Boers had been enjoying some considerable success, in the field, against the British Forces; Ladysmith to be followed by Mafeking, in early 1900, were under siege. The British Army needed urgent and substantial reinforcements, in both men and horses, if the Boers were to be defeated. The North Metropolitan Tramways Co. had an agreement with the Government allowing the War Department to requisition horses for the war and during the period September to November 1899 they selected and purchased, from the company's stud 522 horses at £60 each. The total number of employees 'called to the colours' up to February 1900 had been 166. During 1899 the company had carried about 160,000,000 passengers and were enjoying a period of growth. The drain on their stud could not have come at a worse time, perhaps another good reason to hasten the end of the horse tram era.

An interesting report in Reynolds Newspaper for 9th August 1885 gives us a glimpse into the working practices employed at the time. Under the heading

'The Tramway Slavery' –

> A report on the half yearly meeting of the North Metropolitan Tramways Company, and chaired by George Richardson, there is reference to 'the agitation concerning the hours of work'. The chairman said, the phrase "16 hours per day was calculated to mislead the public into thinking that this was sixteen hours continuous work, whereas one hour and forty minutes were allowed in the middle of the day, and seven minutes on each journey. This reduced the hours of labour to thirteen hours and twenty minutes, the payment being for sixteen hours. It was true that sometimes a servant, owing to the pressure of traffic, did not obtain any relief during the day, but in that case 1s per day extra was paid. They had 273 service conductors and fifty six odd men, and the same number of drivers with a still larger number of odd men". A proposal was put to the meeting by a Mrs Reaney, requesting the directors to "employ a larger number of men, to employ the men only fourteen hours, and to grant every other Sunday off". She said that one man had told her that "during seven months he had only one Sunday off" and another employee had been in the service of the company for thirteen years, had, on average, only two and a half days off in twelve months. The proposal was rejected by the chairman and almost unanimously by the shareholders, only five voting in its favour. There was no mention of the actual wages paid.

In 1897-98 Dick, Kerr and Provincial had joined forces in promoting the Carlisle Tramways and were joined in this by Richardson and Robert B. Barningham, Iron Merchant; Richardson and Barningham were also Directors of the Blackpool and Fleetwood Tramroad Co.

Richardson served as chairman of the Electric Railway and Tramway Carriage Works Ltd. until 1905 and resigned his directorship in 1912.

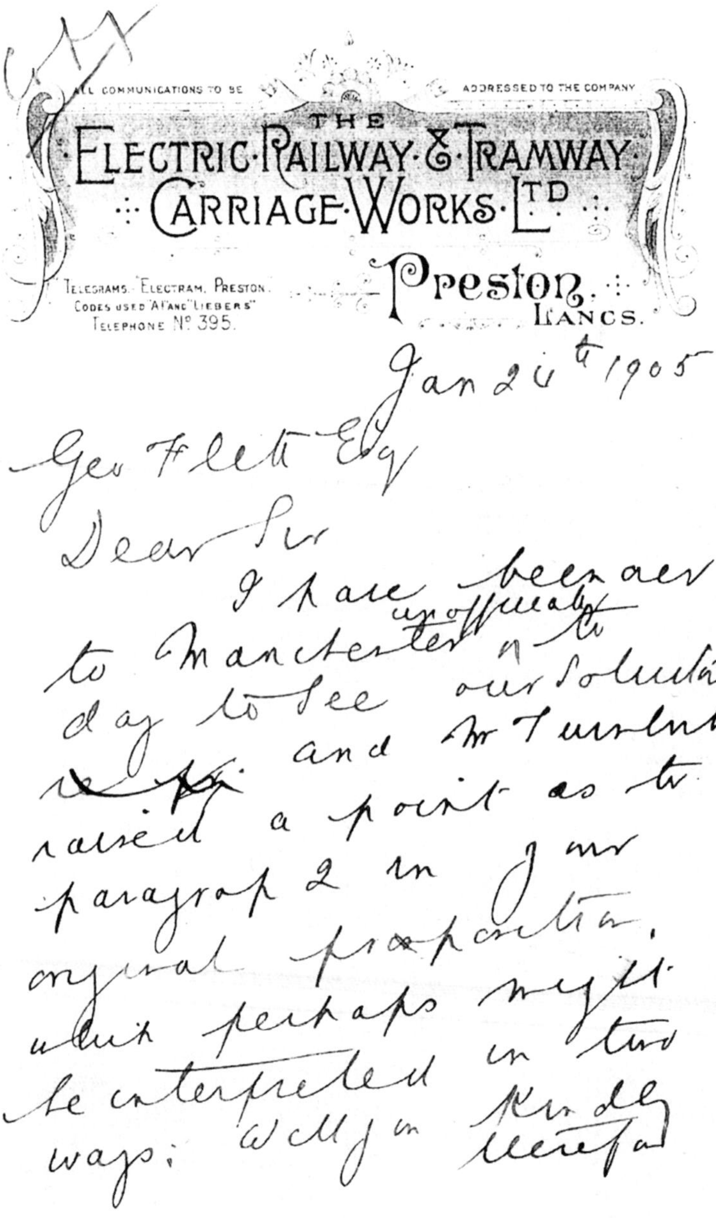

Electric Railway and Tramway Carriage Works. Letter dated 24th January 1905.

The Amalgamation

In a letter to George Richardson dated November 1904 George Flett reports –

I have, in conjunction with Mr. Kerr and Mr. Docker representing the Metropolitan Amalgamated Railway Carriage and Wagon Co. Ltd. for some time past been negotiating for these works and have now acquired the works of Geo. F. Milnes and Co. Ltd. and practically completed negotiations for the acquisition of the British Electric Car Co. Ltd.

I am authorised by Mr. Kerr and Mr. Docker to negotiate with your company (*The Electric Railway and Tramway Carriage Works Ltd*). --- The Metropolitan Amalgamated Railway Carriage and Wagon Co. Ltd. are to have the right to nominate two Directors, Mr. Richardson to remain as Chairman. --- Geo. F. Milnes & Co. properties to be conveyed into the name of Henry Walker, a nominee of Mr. Docker and Mr. Edward Wells, a nominee of myself and Mr. Kerr, to hold as trustees.

On 22nd December 1904 the list of shareholders included the following:

T. Barningham	Strangeways, Manchester	£1365	
R. B. Barningham	41, Victoria Buildings, Manchester	£915	
R. S. Boddington	Strangeways, Manchester	£2910	
W. N. Colam	Waveika, Primrose Bank, Trinity, Edinburgh.	£680	
C. T. Cayley	Constitutional Club, Northumberland Av. S.W.	£1365	
W. B. Dick & Co.	London.	£680	
B. A. Firth	Sheffield	£2730	)
E. M. Firth	Sheffield	£680	) Firth Steel
L. J. Firth	Sheffield	£1000	)
Thomas Firth	Sheffield	£1365	)
George Flett	Abchurch Yard, Cannon St.	£5030	M. D. Dick, Kerr.
G. E. Fry	Dover	£680	
S. B. Fortenbaugh	Underground Electric Railway of London, Hamilton House, Victoria Embankment, London.	£500	
John Kerr MP,	Gaddesden Place, Hemel Hempstead	£1125	
Margaret H. Kerr	Edinburgh	£75	
R. M. Kerr	Edinburgh	£55	
R. H. Prestwich	13,Marsden Square, Manchester	£2000	
J. E. Prestwich	Farnworth	£2000	
G. Richardson	6, Montagu Mansions, Portman Square, London.	£750	
G. Rutherford	(Executors)	£465	
Professor S. H. Short,	(Executors) Cook & Sons, Paris	£1065	
A. Sykes	Preston	£250	
Mrs. A. Sykes	Preston	£125	
B. Sykes	Preston	£50	
C. Sykes	Huddersfield	£250	
T. S. Turnbull	Manchester	Debenture Trustee.	

This shows that W. B. Dick and Co. still retain an interest, be it a small one, otherwise the main shareholders are the Manchester Group, the Kerr family, George Flett and the Firth Steel family of Sheffield, with other company directors and senior employees having a lesser interest.

On 24th December 1904 George Richardson, Chairman requested that the board convene an 'informal and private' meeting of the principal shareholders with regard to the 'desirability of acquiring the two works'; the meeting to be held at the Midland Grand Hotel, Manchester on 6th January 1905.

Messrs Sutton, Elliott, Turnbull and Mayne, Solicitors of Spring Gardens Manchester received an invitation, from the Electric Railway and Tramway Carriage Works, requesting their attendance at a meeting in Manchester on 6th January with a view to 'putting in hand the necessary documents' for the acquisition of G. F. Milnes and the British Electric Car Company.

Shareholders in attendance at the informal meeting held on 6th January 1905, at the Midland Hotel included: G. Flett, J. E. Prestwich, R. H. Prestwich, G. Richardson, R. S. Boddington, R. B. Barningham and J. Kerr.

On 2nd February 1905, in a hand written letter to L. E. Pickin, Robert S. Boddington agrees to a meeting with the British Electric Car Co. Ltd.

On 10th March 1905 from L. E. Pickin to George Richardson, Messrs Prestwich, Barningham and Boddington agree with the proposal that a salary of £1000 be paid to Mr. Milnes for one year.

At an extraordinary General Meeting of the Company on Thursday 18th May 1905 the following resolutions were put to the shareholders:

- Proposed agreement between John Kerr and George Flett, both of Abchurch Yard, Cannon Street, London and William S. Laycock of Oakbrook, Sheffield (the Underwriters) of the first part; the Metropolitan Amalgamated Railway Carriage and Wagon Co., Ltd. of the second part and this Company of the third part, for the sale of the Castle Car Works, Hadley to the Carriage Works. Henry Walker and Edward Wells represented the Castle Car Syndicate Ltd.
- That the name of the Company be henceforth The United Electric Car Co., Ltd.
- The Metropolitan Co. to nominate Lincoln Chandler and Ludford Charles Docker to be Directors, nominated by the Metropolitan Co.

Shareholders of The Electric Railway and Tramway Carriage Works Limited received an invitation to attend an Extraordinary General Meeting of the Company for the purpose of passing resolutions :-

(1) To Authorise the Directors of the Company to enter into two contracts relating to the purchase of the Castle Car Works (late G. F. Milnes) situate at Hadley Salop and the British Electric Car Works at Trafford Park, Manchester and the underwriting and subscription of the shares hereinafter mentioned.

(2) To make certain alterations in the Articles of Association of the Company consequent upon the above.

The Castle Car Works and British Electric Car Works were active Competitors of the Company, but had recently gone into liquidation, and their works were offered for sale, and were purchased jointly by Mr. John Kerr, Mr. George Flett, Mr. William S. Laycock, and the Metropolitan Amalgamated Railway Carriage and Wagon Company, Limited. The Purchasers offered to re-sell both works to the Company for £85,000, which includes a profit of about £20,000 on the original purchase price.

At General Meetings of the Company held on the 18th May and 8th June 1905, resolutions were passed approving the acquisition by the Company of the Castle Car Works (late G. F. Milnes and Co. Limited) at Hadley, Salop, and the British Electric Car Works, at Trafford Park, Manchester, increasing the Capital of the Company, and changing its name to that of the United Electric Car Company Ltd.

The case for the takeover of the two companies was contained in a letter to shareholders dated 1st July 1905, that in the year 1904 there had been a serious drop in the profits of the Company; this, in the opinion of the Directors of the Company, was 'chiefly due to severe competition and cutting of prices'.

'The acquisition of these concerns would not only lessen the competition, but would also enable the Company to largely increase its own output, with a resulting increase in profits'.

As part of the general arrangements under which the two concerns were acquired, two gentlemen connected with the Metropolitan Amalgamated Railway Carriage and Wagon Company, Limited, were nominated Directors of the Company; which Company also took 25,000 of the 100,000 shares on offer.

The business case for amalgamation had previously been presented, at length, by the Chairman, in a speech to shareholders on 18th May 1905, and based on a letter which he had received from Mr. Flett 'as representing Mr. Kerr and Mr. Docker'. The Directors had a number of conferences regarding the proposal and called a meeting of the principal shareholders to get their views on the matter.

> The estimated cost of the two works complete was said to be £65,000, and the original proposition was to sell them to this Company at that price plus £20,000 profit, to be taken in shares; but since Mr. Kerr and Mr. Flett had consented to waive their profit, the only profit on the deal would be that paid to Mr. Docker and the other partner (*believed to have been William S. Laycock a founding director of and a major shareholder in the British Electric Car Co. Ltd.*), i.e. £10,000.

The Metropolitan Company reserved the right to nominate two Directors on the Board; the Chairman considered it would add a "great source of strength to our Company to have two of their Directors on our Board". This was done without increasing the number of Directors, as both Mr. Fry and Mr. Flett resigned from the Board before beginning the negotiations, and they did not intend to seek re-election. Having the Metropolitan Amalgamated Railway Carriage and Wagon Company, Limited, as partners (*they would hold 25% of the shares*) would secure the company from competition with them. In the past the Company had to contend with unfair competition from 'these two bankrupt concerns'; both businesses 'seem to have worked on the most extravagant plans and during the period of their existence have lost irretrievably the whole of their working capital, and within a very short period came into liquidation'. During their existence The Electric Railway and Tramway Carriage Works Limited, 'had to compete with them under very unprofitable circumstances'.

George Flett, on his own responsibility and with the co-operation of John Kerr, and finally also with the 'friendly assistance' of Mr. Docker of the Metropolitan Amalgamated Railway Carriage and Wagon Company, Limited, stepped in and when the two works were in liquidation and in the hands of the court they on their own responsibility bought the works at an 'exceedingly low price' compared with their original cost and did this with the ultimate object of giving to this Company the right of becoming the possessors of them.

Mr. Docker was Chairman of the Metropolitan Amalgamated Railway Carriage and Wagon Company, Limited, the largest concern of its kind in the country; "they had a power of production compared with which ours is a mere nothing". "Mr. Docker had always been on the friendliest terms with Mr. Flett who was a Director of both Companies".

The Metropolitan Amalgamated Railway Carriage and Wagon Company, Limited, nominated Mr. Ludford Docker, brother of the Chairman (*Frank Dudley Docker*) and Mr. Lincoln Chandler who was also an official of their Company.

The agenda book entry for 29th June 1905 contains items:

- To approve date of July 1st for alteration of the company name, the Board Of Trade certificate having been granted.
- G. F. Milnes is on a 2 month salary of £166-13-4.

On 27th July:

- An order from London United for 40 cars.
- G. F. Milnes & Co. and Electric Railway & Tramway Carriage Works – articles of association to be read.

On 21st September 1905, the Memorandum & Articles of Association to the Electric Railway & Tramway Carriage Works Ltd, G. F. Milnes & Co. and the British Electric Car Co. Ltd, also application for a certificate of incorporation to be signed in duplicate by seven subscribers.

On 1st March 1906, the London Office is to review the renewal of Mr. Milnes agreement for one year, since it expired in November.

On 29th March 1906, the board meeting was held at the offices of the Metropolitan Amalgamated Railway Carriage and Wagon Works at Saltley.

This seems to reinforce the significance of the part played by this company in the formation of the United Electric Car Company and its investment in the enterprise.

On 26th April Mr. Stanley tables a report on his recommendation for an extension to the Works.

Customers placing orders, or enquiries, received during 1906/07 (numbers of cars in parentheses) included – Portsmouth (16), East Indian Railway (1+4), Benguella (1), Anglo Argentine, Johannesburg, Alexandria (6), London CC (100/150), London Buses, Bury (6 top covers), Todmorden (4 bus bodies), Porto Alegro (35), Para?(4), Buenos Aires (2), La Capital (125), Delhi (24), South Shields, Kimberley (2), Burton on Trent (3 top covers), London United (40), Rio De Janeiro & Mexican Tramways, Monte Video (30), Calcutta (15), Birmingham (50), Sample Buses for Bellis & Morcom (approval for building), Manchester (20), Blackburn (6), Benguella Railway (3), Rangoon (10), London CC, further enquiry, (300), Sheffield (15), Bury (6) and Burnley (1). This gives a good idea of the scope of work and the international range of the business, the company is clearly not short of work

On 31st January 1907 there is a reference to the repayment of a loan of £1,000 to Mr. Kerr.

Throughout this period there is a considerable amount of correspondence generated between the Carriage Works and Dick, Kerr & Co. on a regular monthly basis concerning orders for cars, presumably with regard to the supply of motors and controllers and possibly Brill trucks; for example:

On 25th July 1907 – debit to Dick, Kerr & Co. - Blackburn £476-6-6, Grand National £615-10-0 and Anglo Argentine £687-19-8.

The correspondence also includes a significant number of letters of complaint with regard to the supply chain.

The United Electric Car Co. came about in August 1905 and had their first board meeting in September of that year the business of the Electric Railway and Tramway Carriage Works continued seamlessly through the handover period. With the exception of George Flett who leaves the board and the addition of Messrs Docker and Chandler from the Metropolitan Amalgamated Railway Carriage and Wagon Co. Ltd. the board members otherwise remains the same. George Flett continues as Managing Director of Dick, Kerr & Co. Ltd. at the London Office and as a director of the Metropolitan Amalgamated Railway Carriage and Wagon Co. Ltd., Saltley.

During the period 1904 to 1914 the Electric Railway and Tramway Carriage Works continues to do business under that name, there is nothing in the records to indicate why this should have been, perhaps as a subsidiary of the United Electric Car Co., correspondence, both sent and received suggests that they operated independently alongside the UEC Co. although under the same board of directors and occupying the same premises.

G. F. Milnes & Co. Ltd., and the British Electric Car Co. Ltd., and the circumstances that caused their decline, and eventual failure as viable companies, is of interest if only to highlight the difficulties being experienced within the tramcar manufacturing industry at the time. The following is a short resume of their histories which may help in understanding why these companies, in particular, should fail as they did and why it was that Dick, Kerr and Co. came to be the predator and they the prey.

A significant source of finance for the undertaking, the Metropolitan Amalgamated Railway Carriage and Wagon Company was not a competitor within the field of tramcar manufacture. The reason for their involvement was in all probability of some significance within the context of their own plans for future growth, if not long term survival.

G. F. Milnes & Co

George Francis Train, an American, was responsible for establishing the first street railway in Britain, opened in Birkenhead in 1860; his trams were assembled locally from imported American parts. About two years later one of his partners, a fellow American, George Starbuck established the first tramcar manufacturing business in Britain at 227 Cleveland-street, Birkenhead. It was established as George Starbuck and Company Ltd. on 12th September 1871 and re-registered one year later as the Starbuck Car and Wagon Company Ltd., on 6th November 1872. In 1878 George Frederick Milnes of Fallowfield, Manchester, became Company Secretary and in 1886 purchased the factory and assets following the winding up of the company. The business was renamed George F. Milnes and Co. and they evolved into a successful tramcar manufacturers, with customers throughout Britain and a substantial export trade.

Tramcar-building in Britain in the mid to late 1890s was a profitable business. Most of the business was in the hands of two specialist companies, G. F. Milnes of Birkenhead and Brush of Loughborough who were able to quote prices which ensured a lucrative return. Rivals in America were unable to compete due to the high cost of shipping.

When, in 1898, the Electric Railway and Tramway Carriage Works Ltd, were established in Preston, followed shortly afterwards, in 1899, by the English Electric Manufacturing Co. Ltd., also on Strand Road in Preston, the 'bespoke' era of tram building came to an end, to be replaced by the mass-produced standard tram, cheap and ready-to-run, and offered at prices which others could not hope to match.

The solution for G. F. Milnes was effectively a take-over bid from the Waggon- und Maschinenfabrik A.G. vormals W.C.F. Busch, of Elmsbütteler Marktplatz 35, Hamburg, who had recently established at Bautzen in Saxony the first large-scale tram building works on the Continent of Europe, and were looking out for business. The Bautzen works was managed by Wilhelm Busch the son of the founder; another of his sons Carl Busch came to England to explore the potential market. Carl Busch was introduced to a Liverpool merchant who had been trying to interest G. F. Milnes in expanding his business. They devised plans for a new car building works, and persuaded Milnes to turn his business into a public limited company. This was registered on 10th September 1898 as G. F. Milnes & Co. Ltd., with a nominal capital of £150,000; G. F. Milnes would be Managing Director of the new company and receive 7½ per cent of the net profit. The other directors, initially, were W. H. Edwards (Chairman), Wilhelm Busch, and Alexander Bruce; the registered office was at 43, Castle Street, Liverpool, the office of the company's solicitor.

The capital required to set up the new works was initially £85,000 of which £58,000 was subscribed by the German firm and its backers W. C. F. Busch, Theodor Horn, Edward Zeitz and S. A. Rosencrantz. The largest British subscribers were Lt. Col. G. H. Morrison, of Liverpool (who would become Chairman), Alexander Bruce, James Taylor and James Jacob Atherton. The site for the new works would be that of the derelict Castle Iron Works at Hadley, near Wellington in Shropshire. Senior staff for the new works being recruited from Milnes at Birkenhead and Busch of Hamburg. Carl Busch was Works Manager at Hadley, assisted by F. Hughes-Caley. The department heads were made up of two recruits from Birkenhead and five from Germany, making it very much a German owned and run operation. Production of tramcars started at Hadley in about June 1900 and the first cars left the factory in late August; these were bogie cars 1-8 for Darwen Corporation.

The period 1901 – 1903 was the busiest for the British tramcar industry. New tramways were being introduced throughout the country; every local authority was aware of the need to provide better and cleaner local transport and the electric tram was the answer. At the beginning of 1901 there were five builders fully employed turning out some 3,000 cars at prices that enabled them to pay a dividend on their shares and the new Hadley works of G. F. Mines had plenty of work turning out some 701 cars. Representatives of the technical press described the Hadley Works as the finest and most modern in the country. In response to a slackening in trade, early in 1902 it was decided to close the Birkenhead works and concentrate all production at Hadley; the registered office and some of the workforce were moved to Hadley in April. Others of the workforce remained in Birkenhead and accepted employment with G. C. Milnes and Thomas Voss at their new works manufacturing tramcar seats and top covers. The rates of pay at Hadley, being in a country district, were less than those in Birkenhead and this resulted in a pay dispute. In August 1902 a delegate conference of the tramcar builders' section of the United Kingdom Society of Coachmakers was held in Manchester to compare wages and conditions and draw up a programme of standard terms and conditions; these were put to the Milnes management in September, and rejected. Since many municipal authorities had adopted the principal of the fair wages clause in their contracts the payment of sub-standard wages could have been held against them when bidding for orders. J. H. Price speculates that this could also be the reason why Milnes failed to bid for some of the bigger contracts and particularly that for London County Council.

From 1902 business is in decline, orders for tramcars are slowing down, capacity within the industry is too great and as a consequence competition is becoming more intense, profits are decreasing and a price war is the inevitable result. The two companies best placed to survive the downturn were the Electric Railway & Tramway Carriage Works in Preston and Brush. The former having financial support from the Dick, Kerr electrical engineering and contracting company with which it had close financial and management connections.

By the middle of 1903 the Castle Car Works was running at a loss with only a small flow of orders at unprofitable prices that could not be sustained. Max Meyer, the Managing Director of the German parent company, came over to assess and report on the situation. His recommendation was that the Milnes business be sold off as soon as possible, before the position deteriorated still further and advised voluntary liquidation. At a special meeting held at Hadley on 18th August 1903, it was resolved to wind up the Company with Max Meyer being appointed Liquidator. The run-down of senior staff was eased by the fact that most of the German engineers and specialists returned to Germany to resume employment with the parent company.

On 3rd-4th September 1903 George Flett and E. A. Stanley, representing Dick, Kerr (George Flett was also a director of the Metropolitan Amalgamated Co.), visited the Hadley Works, at the instruction of the board, to examine the books; hosted by Mr. Meyer and Mr. Milnes. A balance sheet was prepared, dated 31st July 1903, the date the receiver was appointed; the net amount of assets to be taken over amounted to £23,845.

Efforts to find a private buyer for the Hadley Works came to nothing, and early in 1904 it was decided to offer the business for sale by auction. This also came to nothing, the bidding failed to reach a figure acceptable to the auctioneer and the property was withdrawn. The events of 1903-4 had not gone unnoticed among the tramcar manufacturing fraternity.

A syndicate styling themselves the Castle Car Syndicate was formed, comprising John Kerr and George Flett, representing Dick, Kerr and Co. Ltd., the Metropolitan Amalgamated Railway Carriage and Wagon Co. Ltd. and W. S. Laycock, a director of the British Electric Car Co. The Syndicate acquired the Hadley works together with that of the British Electric Car Co. at Trafford Park and in May 1905 offered both businesses as a package to the Electric Railway & Tramway Carriage Works for £85,000.

The Metropolitan Amalgamated Railway Carriage & Wagon Co. subsequently leased the Hadley works from the United Electric Car Co. for an annual rental of £2,500. Metropolitan Amalgamated used the Hadley Works to build railway rolling stock and closed down some of their older and less well equipped properties. For nearly three years the Hadley works produced large numbers of wagons and carriages for the export market. In 1908 a general falling off in trade caused Metropolitan Amalgamated to surrender the lease and in 1910, the UEC Co. and their agent found a buyer for the Hadley works in Joseph Sankey & Co. of Bilston.

Articles of Association were signed in September 1905 and G. F. Milnes & Co. became part of the United Electric Car Co. Ltd. The company was registered as an incorporated company, under the same name, with the UEC Co. continuing to honour outstanding contracts and commitments. The share transfer certificates dated 26th October 1905 name the shareholders each receiving one fully paid £1 share, paid by the United Electric Car Co. Ltd. to be –

George Richardson, Merchant of 6, Montague Mansions, London.	Share No. 1
Richard Henry Prestwich, Cotton Spinner Manufacturer.	Share No. 2
Robert Slater Boddington, Brewer of Strangeways, Manchester.	Share No. 3
John Kerr, Contractor of Abchurch Yard, London.	Share No. 4
Robert Breese Barningham, Iron Merchant of 41,	Share No. 5
Victoria Buildings Manchester.	
Ludford Charles Docker, Manufacturer of Edgbaston, Birmingham.	Share No. 6
Lincoln Chandler, Engineer of Abbotsfield, Kenilworth.	Share No. 7

In 1908 share No. 6 was transferred to Henry Mercer Malcolm of 36, Victoria Street, London.

In 1914 share No. 1 was transferred to Lewis Edwin Pickin of Preston and Share No. 4 was transferred to John Greenwood of The Towers, Brookland, Cheshire.

The first meeting of the Directors of the subsidiary companies (Electric Railway & Tramway Carriage Works Ltd., G. F. Milnes & Co. Ltd. and The British Electric Car Co. Ltd.) was held at the Works, Preston on Thursday 26th October 1905 at 10 am. Present were G. Richardson, R. H. Prestwich, R. B. Barningham and L. C. Docker.

The Secretary reported that the Company had been duly incorporated and the certificate of incorporation was produced, along with a print of the Memorandum & Articles of Association as required. It was resolved that Mr. George Richardson would be the Chairman of the company, Mr. L. E. Pickin, the Secretary and the registered office, the United Electric Car Company Ltd, Strand Road, Preston.

Auditor's reports for G. F. Milnes a Co. continued to be produced annually from 30th June 1906 up to 30th June 1916.

Liabilities	Capital	1,000 shares of £1 each	£1,000
		7 shares of £1 fully paid	£7.00
Assets		the London City & Midland Bank Ltd.	£7.00

Named Directors in 1916 R. B. Barningham and H. M. Malcolm, Secretary L. E. Pickin.

The final meeting of shareholders of G. F. Milnes & Co. was held at Messrs Boddingtons office, Manchester on Friday 9th February 1917 at 3 o'clock. Present were R. B. Barningham (chairman), R. S. Boddington, John Greenwood and L. E. Pickin. A resolution had been passed at an Extraordinary General Meeting of shareholders held at the Cannon Street Hotel, London on Thursday 25th January 1917, that the company be wound up voluntarily and that Mr. L. E. Pickin be appointed liquidator and that the amount of £7-0-0 standing to the credit of the company be transferred to the United Electric Car Company Ltd.

The United Electric Car Company Ltd. was itself taken over by Dick, Kerr & Co. Ltd. in 1917, being the probable reason for the winding up of G. F. Milnes.

The production of tramcars in Birkenhead was continued by George Comer Milnes, the son of George F. Milnes, in partnership with Thomas Voss as Milnes Voss & Co. specialising in the supply of tramcar parts and the fitting of patent top covers to existing open-top trams and manufacturing a small number of complete tramcars. Milnes Voss was wound up in 1913-1914.

The British Electric Car Co. Ltd

With the exception of J. G. Brill of Philadelphia, who had established a London Office at 110 Cannon Street, London, the transatlantic trade in tramcars and their equipment was conducted mainly through agents. Brill had taken over the office space vacated by Dick, Kerr staff when they moved from London to Preston in March 1901. Those agents included Armistead Keith Baylor, an electrical engineer with an office at 18, St. Swithin's Lane, London E. C., and another at 11, Broadway, New York, whose business was described as the purchase and shipment of materials from the USA. Baylor numbered among his associates William Samuel Laycock of Sheffield, a manufacturer of rolling stock accessories such as heaters, ventilators, window gear, blind rollers and all kinds of brassware and castings.

In about 1899 Baylor, being aware of the rapid growth in tramcar manufacturing, decided that there was scope for another tramcar factory in Britain and that he was the man to provide it. Along with Laycock, Baylor also had the backing of William Martin Murphy, chairman of the Dublin United Tramways, who was a tramway promoter and contractor on his own account having imported much American equipment for the Dublin tramways but finding it more economical had progressed to build his own car bodies at Inchinore works, Dublin.

The new company was registered on 2nd August 1900 for the purpose of manufacturing rolling stock for tramways, light railways and ordinary railways.

The site chosen for the new factory was in Trafford Park, Manchester on a plot of land (of 18,502 square yards) leased by Baylor, in 1900, from the Grain Elevator Estates with an option to buy the freehold. More land was bought in 1902, bringing the total to 12 acres. Trafford Park was chosen for the obvious convenience of immediate access to the estate railway, mainline railways and the nearby docks and wharves served by the Manchester Ship Canal.

Investors were slow to support Baylor's view that another tram factory was needed, and raising the capital took longer than intended. To help pay for the factory, a mortgage of £25,000 on the company's land and buildings was obtained early in 1902 from the London Joint Stock Bank Ltd.

The Trafford Park works built its first tram, a sample car completed in October 1901 by which date the new G. F. Milnes plant at Hadley was in full production, Brush at Loughborough had extended their works and these with other builders brought the total number of trams built in Britain in 1901 to more than 3,000. If they were to make any impact and take a share of this business they would need to be competitive on price and delivery. The board recognised this and decided to adopt a strategy of offering quick delivery and quoting prices below those of the older-established firms. This would be achieved by concentrating on a standard design of car, 'lacking the traditional frills', and initially built only for standard gauge track.

The works was operating well below capacity and the spring of 1903 found the BEC Co. short of working capital. The cash crisis was resolved on 17th June 1903 by a £50,000 advance by W. S. Laycock and W. M. Murphy as a mortgage on uncalled capital, soon converted to mortgage debentures on the security of the freehold land which were taken up in part by Trafford Park Estates Ltd., A. J. Smith, and the Grain Elevator Estates Ltd.

Production during the first twelve months was only about one quarter the capacity of the works and by the end of 1903 the total output was only about 300 cars. Their prices were competitive but contracts were not always awarded to the lowest bidder, reputation and previous performance was also important to the customer.

The company then decided that the way forward was to cut profit margins and go in for deliberate price-cutting. This only served to invite retaliation from other manufacturers including G. F. Milnes & Co., which as we have already seen, was driven into liquidation in August 1903. The next victim was BEC itself, whose turn came three months later.

By November 1903 it was clear to the shareholders that tramcar building was not the way to the quick fortune that Baylor had promised them and decided to put the firm into liquidation, trading was ceased immediately. An Extraordinary General Meeting of shareholders was held at the works on Monday 7th December 1903, at which it was resolved 'That it has been proved to the satisfaction of this meeting that the Company cannot by reason of its liabilities continue in business and that it is desirable to wind up the same'. T. F. Wild's appointment as liquidator was confirmed, and on 21st December the company's registered office in London was transferred to his chambers at Broad Street Avenue.

On 8th June 1904 the adjourned meeting of the creditors of the British Electric Car Co. Ltd. (in liquidation) was held at the Mosley Hotel, Manchester, Mr. T. F. Wild (liquidator) presiding.

The Chairman reported that efforts had been made to sell the business as a going concern, but without success. After much discussion it was resolved that 'the matter remain in the hands of a committee of creditors with the request that they formulate a scheme for the purchase of the business and ascertain, by way of letter, the willingness or otherwise of the creditors to subscribe thereto, and if sufficient support be not obtained, the scheme be abandoned'.

When, in 1904, BEC had failed to attract a buyer, W. S. Laycock took the decision to join forces with the Castle Car Syndicate and persuaded them to acquire the Trafford Park works along with the Hadley works of G. F. Milnes on behalf of the Electric Railway & Tramway Carriage Works Ltd. This was an attractive proposition for Dick, Kerr and Metropolitan Amalgamated since it would ensure the properties would not fall into the hands of others in the industry.

L. E. Pickin Secretary of the Electric Railway & Tramway Carriage Works Ltd receives a letter dated 2nd February 1905 from Robert S. Boddington, Boddington Breweries Ltd., Strangeways agreeing to meet the chairman of the British Electric Car Co. at the works. On 12th April 1905 George Richardson, chairman of the Electric Railway & Tramway Carriage Works Ltd., receives notification from the company solicitor, Messrs. Sutton, Elliott, Turnbull & Mayne of 13 Spring Gardens, Manchester, that 'the court has confirmed the contract for the sale of the Trafford Car Works'. The purchase was agreed in May 1905 for £85,000 including a profit of £20,000 for the members of the Syndicate, and the two businesses became incorporated along with the Electric Railway & Tramway Carriage Works to create the United Electric Car Co. Ltd. in Preston.

The only person to emerge unscathed was A. K. Baylor, who while urging others to invest in BEC had failed to provide any capital of his own. While the other directors were busy trying to rescue something from the mess, Baylor had become the UK agent for the American Car & Foundry Company of Berwick, Pennsylvania and the supply of 108 all steel tube railway cars for London's Bakerloo line, which ironically were assembled in the UK on a site at Trafford Park only 200 yards from the now abandoned BEC works.

The lease on the Trafford Park works remained with the United Electric Car Co., empty and disused until 1911 when transfer of ownership becomes an issue and the United Electric Car Co. was being sued for rent-charge; correspondence with the company Solicitor provides confirmation of previous conveyances.

- Land on which these premises formed part was conveyed (9th October 1902) by the Trafford Park Estates Co. Ltd. and their trustees at rent-charge of £631-17-0 to the Grain Elevator Co. Ltd.
- Then, 31st December 1902 to the British Electric Car Co. Ltd., conveyed by the Elevator Company.
- Then, December 1903 the Car Company wound up and by indenture of 1st August 1905 conveyed by the Car Company and their debenture Trustees and Liquidators to the Castle Car Syndicate Ltd.
- Then, by indenture dated 1st August 1905 conveyed by the Syndicate to the United Electric Car Co. Ltd.
- The Trustees authorised to proceed with the conveyance of the premises, dated 3rd April 1911.

The Company held on to the Hadley and Trafford Park premises, for during the period 1909-1910 there is correspondence between L. E. Pickin, for the Company, and the Metropolitan Amalgamated Railway Carriage and Wagon Company, Ltd. on the subject of rates for the period 1908-1909 on the Hadley Works and with a London Auctioneer & Estate Agent on the sale of the Trafford Park Works. A handwritten note, dated 6th May 1910 from Robert B Barningham at Harrogate to Mr. Pickin makes reference to talks taking place with Marshal Stevens, Trafford Park Estates Manager, of Manchester "who are trying to get together a syndicate to start a wagon works for the manufacture of wagons for the ship canal company".

On 16th August 1911 the lease on the BEC works and the adjoining car shed site was acquired by the Ford Motor Company to build the first Ford cars outside America, the model 'T'.

The Metropolitan Railway Carriage and Wagon Company Ltd

Joseph Wright was a London coach builder in the early part of the 19th century and was a contractor for the Royal Mails and the owner of most of the stage coaches then running between London and Birmingham, he was the founder of the Metropolitan Company. During the early 1840s he realised that the era of the stagecoach was drawing to a close and he began to devote his energies to the manufacture in his London works of railway carriages. As early as 1844 he obtained a very wide patent (No. 10173) in respect of improvements in railway carriages, detailed carriages with 4,6 and 8-wheeled bogies. As the railways grew the demand for rolling stock was such that Joseph Wright decided to build a new factory outside London where he would have unlimited space for expansion. Birmingham was already the junction of a number of railways and an area that he knew well, from his earlier coaching activities, and with its central position and proximity to coal and iron districts, it was an attractive location. He decided, therefore to build his factory in this area and in March 1845 was able to lease some 6 acres of land at Saltley.

He was assisted in the business by his sons Henry and Joseph and traded as Joseph Wright and Sons, Railway Carriage and Wagon Builders and Contractors of Saltley Works, Birmingham, with a London office at 8, Adam Street, Adelphi. By 1847 the business had prospered and Wright obtained leasehold on adjoining land at Saltley covering substantially the remainder of what was to become the Saltley Works site of approximately 50 acres. In 1853 the original works, known as the Saltley Old Works, were let to the London and North Western Railway Company whilst Joseph Wright and Sons continued in the adjoining new works, employing about 800 people. After the death of Joseph Wright, in 1859, the business was continued by his sons under its original name until 1862 when the Metropolitan Railway Carriage and Wagon Company Ltd., was formed with a nominal share capital of £100,000.

In 1902 the first step in the rationalisation of the rolling stock industry was made when the Metropolitan Amalgamated Railway Carriage and Wagon Company Ltd., was formed with a share capital of £1,500,000 to take over the business of the Metropolitan Company and to amalgamate with the following well known rolling stock companies:-

- Ashbury Railway Carriage and Iron Co.
- Brown Marshalls and Co.
- Lancaster Railway Carriage and Wagon Co.
- Oldbury Railway Carriage and Wagon Co.

In the same year, in order that they might have their own source of supplies for the building of rolling stock, the amalgamated company acquired the whole of the assets and undertaking of the Patent Shaft and Axletree Company Ltd. of Wednesbury, an important manufacturer of iron and steel with their own rolling mills. In 1905 the Metropolitan Amalgamated Railway Carriage and Wagon Company Ltd. part financed, along with John Kerr and George Flett, the formation of the United Electric Car Co., holding about 25% of the shares and placing two of their Directors, Ludford Charles Docker (brother of the Chairman, Dudley Docker) and Lincoln Chandler, on the board of the new company, George Flett was already a director of both companies and also a director of the Daimler Motor Co. of Coventry along with Dudley Docker and Lincoln Chandler.

In 1910 the Institution of Mechanical Engineers visited the Saltley works and reported on the visit in some detail, with a description of the site, its location and manufacturing processes. The works, at the time, covered about 43 acres and employed about 3,000 men.

In 1919 Vickers Ltd. acquired the shares of the Metropolitan Company.

Petrol Railcars

During the period leading up to the merger of the three companies and founding of the United Electric Car Company it would appear that Dick, Kerr were taking an interest in the development of cars powered by the internal combustion engine. These took the form of railcars with petrol-electric drive or with direct drive from a petrol engine. This would appear to be a natural progression since sales of electric tramcars were falling and rapid advances were being made in the development of the internal combustion engine which, at the time, would have made it the logical successor as the alternative means of traction, particularly for use on light railways.

The Sheppey Light Railway

In September 1903, 'due to an extreme decrease in passengers during the last year' it was proposed to replace the steam locomotives with railcars, or rail-motors 'as they were sometimes known'. Preliminary runs were made on the Sheppey Light Railway with two petrol-electric railcars that were borrowed from Dick, Kerr & Co. These were not very successful, especially on frosty mornings when the 'familiar ignition problems with internal combustion engines were experienced'.

Great Northern Railway Company

The Engineer journal of 21st July 1905 reports –

> The North-Eastern Railway Company have recently adopted petrol for locomotive purposes, but they have complicated it by the addition of an electric dynamo. How far this will be successful remains to be seen, but it is certain that the weight of the petrol electric car is as great as the steam motor car. The cost of construction is also more, and the cost of upkeep will certainly not be less. To its credit it must be said

> that it dispenses with the use of one man, which is equal to the capitalised value of £1,200. The Great Northern Railway Company have gone a step further, and have adopted the petrol engine to directly drive the car itself.

The Great Northern Railway had many small branch lines running through thinly populated districts, which though important as feeders of the main lines, were often worked at a loss. An 'automobile carriage' was identified as the most likely solution to the problem with the added benefit of being able to 'prevent the competition of road motors'. Messrs Dick, Kerr and Company had built a car, which, when fully loaded weighed about sixteen tons. The motive power was supplied by two four-cylinder petrol engines of the Daimler type, each giving 36 hp. The engines were suspended on a 'special frame' independent of that which carried the coach body, so that 'the passengers are not troubled by the vibration of the engines'. The engines developed sufficient power to drive the car at 50 mph and after travelling 2,000 miles the 'machinery' showed no appreciable signs of wear. The company considered that it had found 'a reliable type of motor for the working of the short line between Hatfield and Hertford'.

The fitting of a Daimler engine to a car body manufactured by the United Electric Car Company is a logical partnership since George Flett, managing director of Dick, Kerr and one of the chief architects of the founding of the UEC Co., with Ludford Charles Docker and Lincoln Chandler, directors of the UEC Co., were all directors of the Daimler Motor Co.

London, Brighton and South Coast Railway

The Engineer journal of 25th August 1905 reports –

> That the petrol tram car can be used with success is demonstrated by the working of those in the Liége Exhibition grounds. --- It is a suggestive fact that Messrs. Dick, Kerr and Co., have just delivered two petrol railway motor cars to the London and Brighton Company.

The London, Brighton and South Coast Railway had become concerned about loss of revenue on branch and short-distance passenger services, particularly in winter. The directors asked their Chief Mechanical Engineer, Robert Billinton to investigate the use of steam or petrol railcars on lightly used services. Billinton died in 1905 to be succeeded by Douglas Earle Marsh who acquired two steam railcars from Beyer, Peacock and Company and two petrol railcars from Dick, Kerr & Co. These were compared with small steam locomotives of the Stroudley A1 and D1 classes fitted for 'motor train' or 'push-pull' working. Neither type of railcar was successful, being inadequate to cope with traffic fluctuations between winter and summer.

Petrol Railcar. One of two Petrol Railcars built by Dick, Kerr & Co. in1905 for the London, Brighton and South Coast Railway and operating between Lewes and Seaford.

The United Electric Car Company Ltd.

(Incorporated in 1898 as The Electric Railway and Tramway Carriage Works Ltd):

The London office of the United Electric Car Company, formerly the Electric Railway & Tramway Carriage Works Ltd was at River Plate House, 13. South Place E.C. with the Head Office in Preston.

On 6th April 1905 a letter to George Richardson, Chairman of United Electric Car Company Limited, from Mr Walker of the Castle Car Syndicate with reference to the take-over of the Hadley Works to 'give notice to all employees and clear them out'. This was also copied to the Metropolitan Amalgamated Railway Carriage and Wagon Company, Ltd., Saltley.

On 26th July Ludford Docker receives notice from H. Walker that the Castle Car Syndicate requires £85,000. The correspondence with Ludford Docker was, presumably, because he represented Metropolitan Amalgamated Railway Carriage and Wagon Company on the board of United Electric Car Company Limited being the largest single shareholder in the new company.

On 12th August 1905 the Company Secretary (L. E. Pickin) receives from the Company Solicitors, Messrs Sutton, Elliott, Turnbull and Mayne of Spring Gardens, Manchester the Memo of Association of each of the Companies, Geo. F. Milnes & Co. Ltd., the British Electric Car Co. Ltd. and the Electric Railway and Tramway Carriage Works Ltd. also, the Memo of Articles of Association of the United Electric Car Company Limited.

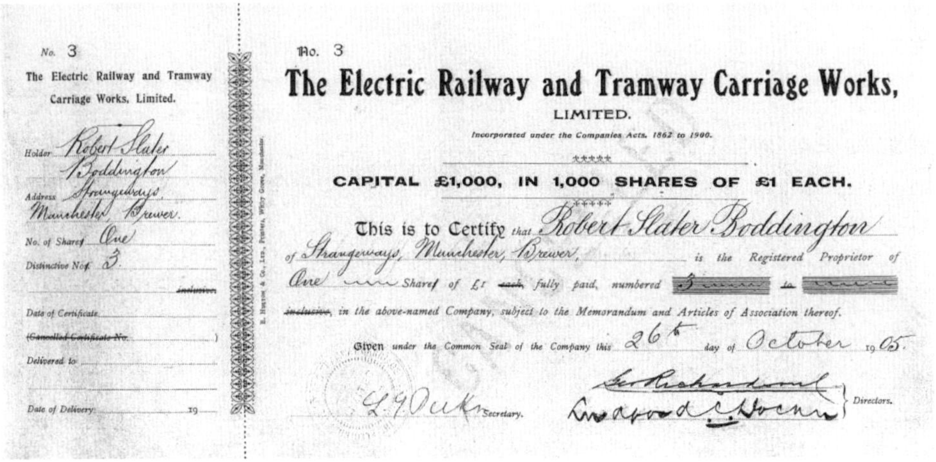

No. 3

The Electric Railway and Tramway Carriage Works, Limited.

Holder Robert Slater Boddington

Address Strangeways Manchester, Brewer.

No. of Shares One

Distinctive Nos. 3

~~inclusive~~

Date of Certificate

~~(Cancelled Certificate No.~~)

Delivered to

Date of Delivery 19

No. 3

The Electric Railway and Tramway Carriage Works,

LIMITED.

Incorporated under the Companies Acts. 1862 to 1900.

*****

CAPITAL £1,000, IN 1,000 SHARES OF £1 EACH.

*****

This is to Certify that Robert Slater Boddington of Strangeways, Manchester, Brewer, is the Registered Proprietor of One Share of £1 ~~each~~ fully paid, numbered 3 ~~to~~ ~~inclusive~~, in the above-named Company, subject to the Memorandum and Articles of Association thereof.

Given under the Common Seal of the Company this 26th day of October 1905.

Secretary.

Directors.

The Electric Railway and Tramway Carriage Works Limited. Share Certificate dated 26th October 1905 in the name of Robert Slater Boddington.

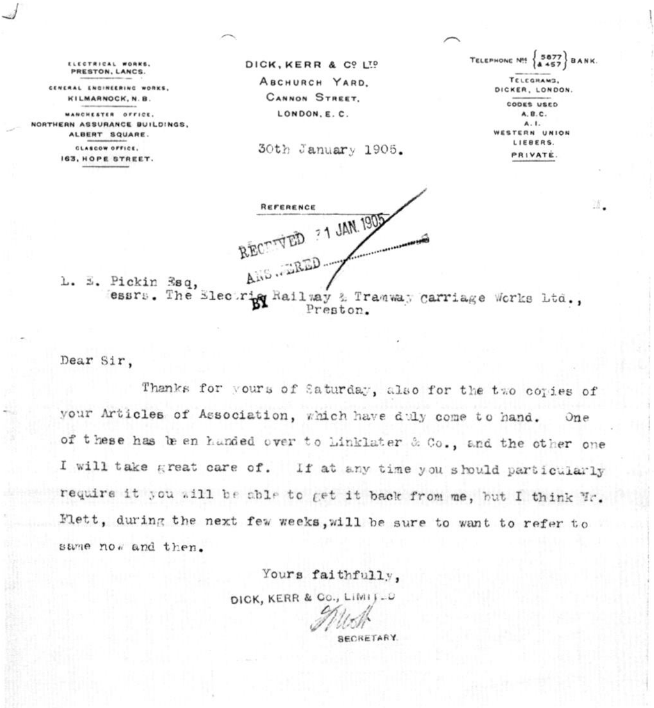

ELECTRICAL WORKS, PRESTON, LANCS.
GENERAL ENGINEERING WORKS, KILMARNOCK, N. B.
MANCHESTER OFFICE, NORTHERN ASSURANCE BUILDINGS, ALBERT SQUARE.
GLASGOW OFFICE, 163, HOPE STREET.

DICK, KERR & Co LTD
ABCHURCH YARD,
CANNON STREET,
LONDON, E. C.

TELEPHONE NOS {5877 & 457} BANK.
TELEGRAMS, DICKER, LONDON.
CODES USED A.B.C. A.I. WESTERN UNION LIEBERS.
PRIVATE.

30th January 1905.

REFERENCE

RECEIVED 31 JAN. 1905
ANSWERED
BY

L. E. Pickin Esq,
Messrs. The Electric Railway & Tramway Carriage Works Ltd.,
Preston.

Dear Sir,

Thanks for yours of Saturday, also for the two copies of your Articles of Association, which have duly come to hand. One of these has been handed over to Linklater & Co., and the other one I will take great care of. If at any time you should particularly require it you will be able to get it back from me, but I think Mr. Flett, during the next few weeks, will be sure to want to refer to same now and then.

Yours faithfully,
DICK, KERR & Co., LIMITED
SECRETARY.

Letter from Dick, Kerr and Company Limited. To the Electric Railway and Tramway Carriage Works Limited. Acknowledging receipt of their Articles of Association. 30th January 1905.

List of Shareholders present at the informal meeting of the principal Shareholders held at the Midland Hotel, Manchester, on Friday the 6th January 1905.

-:-:-:-:-:-:-:-

		No. of shares.		£	
	G. Flett	1279		6395	
In Bankers hands.	B. Sykes	753		3765	
	A. Westworth	546		2730	
do.	G. Readman & others	2560		12800	
	F. W. Smith	226		1130	
	C. H. Rannard	370		1850	
	D. Sanderson	230		1150	
	J. E. Prestwich	400		2000	
	F. J. K. Cross	500		2500	
	M. B. Copland	273		1365	
	J. E. Jones	273		1365	
	J. R. Hoyle	200		1000	
	O. O. Wrigley	400		2000	
	Mrs. Bruce	100		500	
	F. Wrigley	273		1365	
	P. R. Crompton	136		680	
	J. S. Walker	50	8569	250	42845
	DIRECTORS:-				
	G. Richardson	~~100~~ 150		~~500~~ 750	
	R. S. Boddington	582		2910	
	R. B. Barningham	183		915	
	R. H. Prestwich	400		2000	
£500 in Bankers hands.	J. Kerr	325	1590	1625	7950
			10159		£ 50795

NOTE:- In addition to the above, Mr. D. Paulin, Edinburgh, was also present. He is not a shareholder in our books but stated re represented some Scotch shareholders and we are writing him today to know whom he represented.

23rd May 1905.

See Mr Paulin's letter rec'd 25/5/05 — Clydesdale BK Edinburgh 1022 shares £5110; do London 546 £2730

List of Shareholder. At the informal meeting held at the Midland Hotel, Manchester on Friday 6th January 1905.

H.H. S. DEPT:

CODES USED, "A.1" "LIEBER'S" AND "WESTERN UNION".

ALL COMMUNICATIONS TO BE ADDRESSED TO THE COMPANY.

TELEGRAMS:- "PRESCAR, LONDON." LONDON OFFICE, RIVER PLATE HOUSE, 13, SOUTH PLACE, E.C. TELEPHONE Nº 458 LONDON WALL.

The United Electric Car Co. Ltd.

(FORMERLY THE ELECTRIC RAILWAY & TRAMWAY CARRIAGE WORKS LTD)

INCORPORATING

GEORGE F. MILNES & CO. LTD., HADLEY. THE BRITISH ELECTRIC CAR CO. LTD., MANCHESTER.

TELEGRAMS:- "PRESCAR, PRESTON." TELEPHONE Nº 395.

REGISTERED OFFICES:- PRESTON, LANCASHIRE.

Head Office, Preston, Lancs.

2nd July 1906.

The Trafford Park Estates, Ltd.
Trafford Park,
MANCHESTER.

Dear Sirs,

Will you please let us have your receipt for cheque for £2. 10. 0. sent you on May 30th.

Yours faithfully,

THE UNITED ELECTRIC CAR CO., LTD.

SECRETARY.

United Electric Car Company Limited. Letter from 1906.

Tramcar Assembly. The United Electric Car Company Ltd., Strand Road.

Tramcar Manufacturing at the Works of The United Electric Car Company Ltd. Wood Working Department.

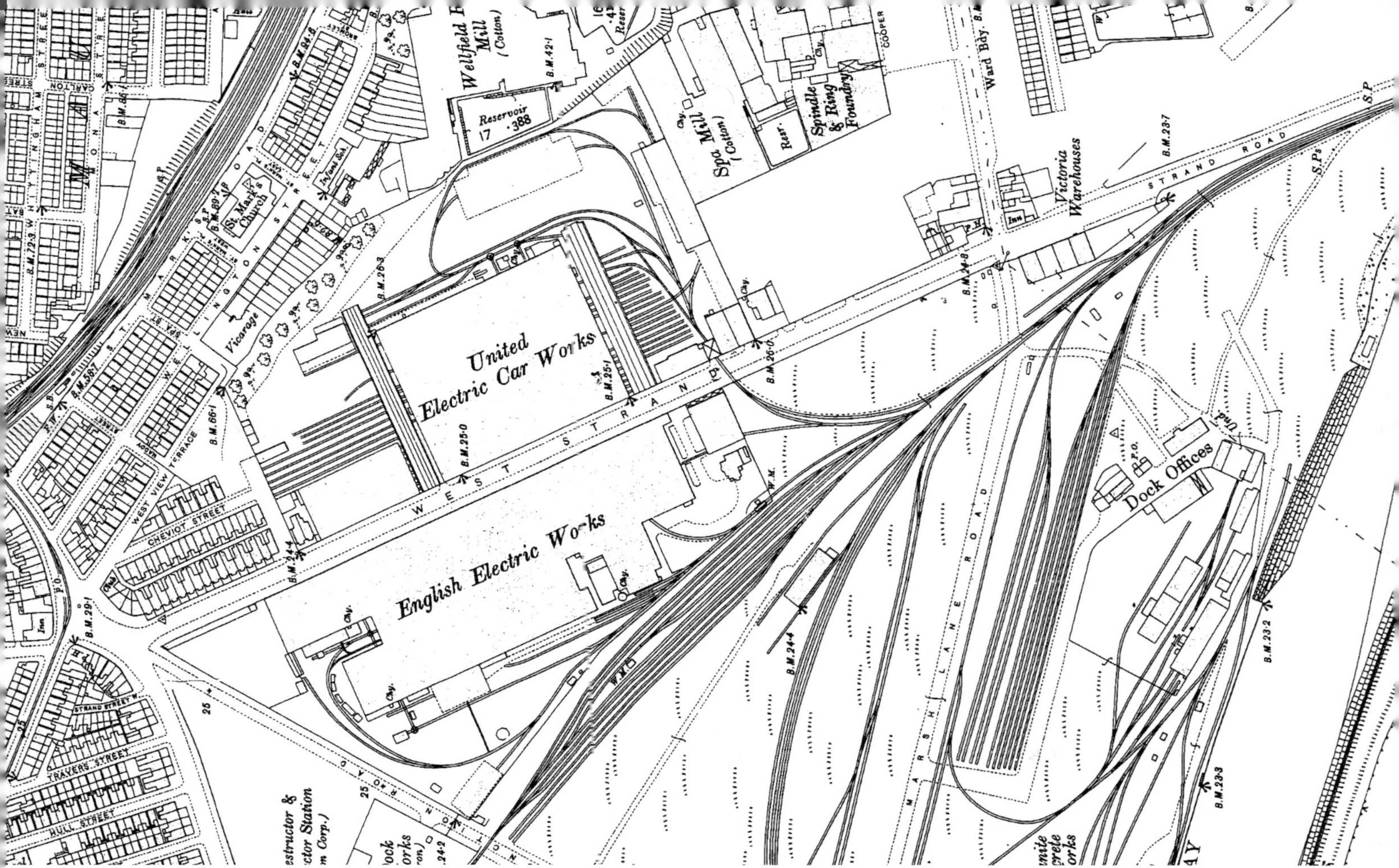

Early Map of Strand Road after the formation of The United Electric Car Company in 1905. Clearly showing the Main Line Railway Connections and Tracks into and out of the Tram Works.

On 20th September the Company Secretary is informed that the United Electric Car Company Limited is not entitled to the books of the British Electric Car Co. Ltd. except as a matter of favour until the winding up has been completed. George Flett responded by offering to take the matter up personally.

On 22nd November, L. E. Pickin, Secretary, gave notice to the three companies of a statutory meeting, required to be held by section 12 of the Company's Act of 1900, to be held at the offices of the United Electric Car Company Ltd., Strand Road, Preston.

The first Directors meeting of the newly created United Electric Car Company Limited, on 30th September 1905 had the following attendees:

G. Richardson,	Chairman	Lombard Court, Lombard Square, London.
R. H Prestwich		13 Marsden Square, Manchester.
John Kerr M.P.		Loudwater, Rickmansworth, and Gaddesden Place, Hemel Hempstead, Herts.
R. S. Boddington		Strangeways Brewery, Manchester.
R. B. Barningham		41 Victoria Buildings, Manchester & St. Trinians, Kent Road, Harrogate.
L. C. Docker		Edgbaston, Birmingham.

Lincoln Chandler attended some later meetings.

Over the next few years, including the period of the Great War, the venue for some of the Board Meetings was Boddingtons Brewery in Manchester.

The first list of shareholders included, in addition to the above named Directors:

G. Flett	Abchurch Yard, Cannon Street, London.
J. E. Prestwich	Farnworth, near Bolton.
C. T. Cayley	Constitution Club, Northumberland Avenue, W.C.
Thomas Barningham	Strangeways, Manchester.

On 8th June 1905 the Hadley Works is leased to the Metropolitan Amalgamated Railway Carriage and Wagon Co. Ltd, Amalgamated to pay £2,500 rental.

Also dated 26th October 1905, an agreement is signed with the City of Manchester, for the supply of water to the United Electric Car Company Limited of Preston, signed by the Lord Mayor and Town Clerk for Manchester and Geo. Richardson, Ludford C. Docker and L. E. Pickin for the company.

Agenda book No. 3 for the period March 1907 to March 1910, gives details of further orders and enquiries for cars, including –

Chester (5), Alexandria (6), Rotherham (9), Stockport (5), Douglas (2 horse cars & 2 cable cars), Buenos Aries & Belgrano (15), Monte Video (70), Doncaster (4), Anglo Argentine (50), Salford (a water cart), Athens (100), Luton (12), Daimler Motor Co. (1 bus body), Johannesburg (100), Buenos Aires & Grand National (125), Burma (3), Westinghouse Co. (1 metre gauge coach), Singapore (6), Bakia (6), Maidstone (2),

Para. (29) through J. G. White & Co. and (10) through Dick, Kerr & Co., Stockport (10); Birmingham (30), and Singapore (4) also through Dick, Kerr & Co. Lanarkshire (7), Blackburn (6), Dumbarton (20) through British T. H. Co., India (8), Greenock (1), Rotherham (3 conversions), Swansea (2 top covers) and Southport (2 top covers).

On 25th April 1907 it is proposed that George Milnes remain on salary at £83-6-8.

On 25th July it is proposed that John Kerr be paid the balance on a loan with interest £1,113-14-0.

Directors expenses declared at the same meeting provide us with a current list of board members –

George Richardson (chair), Richard Henry Prestwich (cotton spinner), John Kerr, Robert Breese Barningham (iron & steel merchant), Robert Slater Boddington (brewer), Ludford Charles Docker, Lincoln Chandler, Thomas Strover Turnbull (Solicitor) and James Edward Prestwich (cotton spinner).

From each meeting there are three suppliers in particular who are singled out for payment of their bills, these are, for example on 21st January 1909 – Docker Brothers (£84-14-8), Gabriel & Co. (£180-5-0) and Brill & Co. (£487-15-0).

The orders and enquiries continue to flow in and for the period 1908 to mid 1909 show some interesting variations from the previous stream of orders for trams, including –

A new order from London County Council for 175 cars and trucks; an electric locomotive for Dundee via British Thompson Houston Co; Char-a-bancs for Renard Road & Rail Transport Corporation Ltd and 4 ambulance cars for the Metropolitan Asylums Board.

On 2nd July 1908 Henry Mercer Malcolm is nominated to be the replacement for Ludford Docker representing the Metropolitan Amalgamated Railway Carriage & Wagon Co. on the board; he attends his first meeting on 26th August 1908.

On 29th September 1909, and again in 1913, the company records (Electric Railway & Tramway Carriage Works) give details of a return under the Companies (Consolidation) Act of 1908 (form E) a summary of share capital and shares of £1,000 in 1,000 shares of £1 each allocated to named directors:

George Richardson	6, Montagu Mansions, London S.W.	Merchant.
Richard Henry Prestwich	13, Marsden Square, Manchester.	Cotton Spinner.
Robert Slater Boddington	Strangeways, Manchester.	Brewer
John Kerr	Abchurch Yard, London E.C. and Loudwater, Rickmansworth, Herts.	Contractor
Robert Breese Barningham	41, Victoria Buildings, Manchester.	Iron Merchant.
Lincoln Chandler	Abbotsfield, Kenilworth.	Engineer.
Henry Mercer Malcolm	36, Victoria Street, London S.W.	

Signed - L. E. Pickin, Secretary.

The list of expense accounts confirms Messrs Turnbull and Prestwich as debenture Trustees.

On 17th March 1910 a patent agreement, with John E. Anger, is signed and sealed, and on 21st December Mr. Alan S. King is appointed representative at the London Office.

Agenda book No. 4 for the period 1910 to 1915, shows that the car business continues as before with work for the war effort starting to take effect –

Subscription to the National Relief Fund £100-0-0, Orders from the War Office for 2,500 cartridge boxes and 72 timbered wagons. Mr. Angers reports on his visits to Farnborough, London and Woolwich, perhaps the first steps leading up to future war related work. Orders for cars continue: Blackburn (4 top covers), Anglo Argentine (700 'triumph' seats), Colne (5 top covers), Dumbarton (2 top covers), it would appear that operators of the company's trams are covering the top decks of their open top trams.

On 15th April 1911 Thomas Croft & Sons of Victoria Street, Preston provides an estimate for the erection of an additional bay to the works similar to that at the north end, £1,487-10-0.

Throughout the period 1907 to 1912 the venue for board meetings was, under normal circumstances, at the Preston Works. Occasionally meetings would be held at other sites, including Boddingtons Brewery, the Metropolitan Amalgamated Railway Carriage & Wagon Co. (later the Metropolitan Amalgamated Railway Carriage & Finance Co.), Saltley Works, the Hadley Works or the offices of Sutton, Elliott, Turnbull & Mayne, the company Solicitors.

On 11th June 1912 a land valuation for the Castle Car Works and land at Hadley of 13 acres, 0 roods, 33 perches and occupied by the United Electric Car Co. Ltd, including houses & gardens in Trench Road, Hadley gave an original total value of £22,900-0-0.

From 1914 the Company used surplus capital for loans to various businesses and local authorities.

On 6th May 1914, a letter from Lincoln Chandler representing the Metropolitan Amalgamated Railway Carriage and Wagon Co. Ltd. of Saltley, Birmingham to L. E. Pickin of the United Electric Car Company Ltd. concerns the loan of £10,000 to Messrs Bolckow Vaughan & Co., Middlesbrough. The exchange of correspondence involved Erasmus Darwin (the grandson of Charles Darwin) in his capacity as a director of that company. Initially R. B. Barningham, Chairman, would not agree so the decision was postponed until a meeting on 28th May. In July the loan was agreed at interest of 4% pa.

Further loans made over the period 1914-17 include –

The Globe Cotton Spinning & Manufacturing Co. Ltd.	£5,000-0-0 at 4%.
Town Hall, Salford	£10,000-0-0 at 5%
Town Hall, Preston	£5,000-0-0
Bury & District Joint Water Board	£10,000-0-0

John Kerr retired in 1912 and the General meeting of 18th September 1912 was his final meeting; he continued to be a shareholder in the business; the register of share transfers for the period for June 1912 shows him transferring 1,125 shares whilst in the period September – October 1912 some 1,025 shares are transferred to him, 75 of these from his wife. In August 1912 Dorothy Maude Kerr of Regents Park, London acquires 2,000 shares from W. Rutherford and Frank Mott of Abchurch Yard, Cannon Street, Directors of Dick, Kerr.

On 25th October 1916 the Secretary writes to Brush asking 'how much they are paying in wages, Body-makers, Finishers, Painters, Engineers, Tinsmiths, Woodworkers and Labourers, in confidence'. The letter was also copied to Messrs. Hurst Nelson and Co. and to Lincoln Chandler at Saltley.

The reply is received on 28th October –

Trade	Pre-War	Present
Fitters/Turners	38/-	43/-
Blacksmiths	38/-	43/-
Strikers	25/-	30/-
Tinsmiths	38/-	41/-
Wood Machinists	40/6d	40/6d + 2/- War Bonds
Body-makers	39/6d	39/6d + 2/- War Bonds
Cabinet Makers & Finishers	37/6d	37/6d + 2/- War Bonds
Coach Painters	36/10d	36/10d + 2/- War Bonds
Labourers	23/-	28/-

On 15th November 1916 the Chairman requests a meeting with Messrs Malcolm, Greenwood, Prestwich and Boddington to discuss an offer he has received from Messrs Dick, Kerr and Company "To purchase all the shares in our Company"

On 16th November, a report is sent to Sir W. B. Peat and Co. at 11, Ironmongers Lane, London with reference to the requirements of the Board of Trade Committee on Engineering Industries (*no doubt in connection with the above*):

Amount of sales	1912	£58,426-19-0 home,	£117,122-13-0 export.
	1913	£75,908-3-8 home,	£59,680-7-8 export.

Value of plant and machinery 1913, £50,616-18-2 excluding land and buildings.

On 23rd November, W. Rutherford of Dick, Kerr and Co., Abchurch Yard, EC was informed that lists of Ordinary and Preference shareholders with names and addresses was being prepared.

A letter is received from Dick, Kerr's Solicitor, dated 30th November 1916 –

'It will be necessary that you're Directors, other than Mr Prestwich, should retire, and that as they retire they should elect nominees of our clients'. Then with reference to a board meeting set for 9th December – 'Our clients (Dick, Kerr) would of course, at this meeting, take over and pay for the shares held by the directors and would also pay the agreed compensation'.

Note - Richard Henry Prestwich would become deputy chairman of Dick, Kerr.

Fusion of Two Engineering Companies

(Times 15th December 1916)

Dick, Kerr and United Electric Car

> A fusion will shortly take place, our Preston Correspondent states, between two large companies which have extensive works in the Strand-road, Preston - Messrs. Dick, Kerr and Co. and the United Electric Car Company. During the past few days notices have been issued to the shareholders of both companies acquainting them with the scheme.
>
> The directors of the United Electric Car Company state that they have been approached by Messrs. Dick, Kerr and Co. who desire to acquire all or as many as possible of the shares in this company. As a result of negotiations they had agreed to transfer to Messrs. Dick, Kerr and Co. the shares held by them and to retire as directors upon condition that every shareholder shall be given an opportunity of transferring his shares to Messrs. Dick, Kerr and Co. on the same terms as themselves – namely, for each preference share in the Electric Car Company one fully-paid Six per cent Cumulative Preference Share of £1 in Messrs. Dick, Kerr and Co., carrying the full dividend due on April 1, 1917, or at the option of the shareholder 20s in cash.

In 1917 the United Electric Car Co. was taken over by Dick, Kerr and Co. Ltd.

The liquidation of the three companies, making up the United Electric Car Company Ltd., continued to generate correspondence.

On 10th February 1917, to the London City and Midland Bank Ltd., Furthergate, Preston, with reference to subsidiary companies: Transfer the sum of £7-0-0 at present standing to the credit of each – Electric Railway and Tramway Carriage Works Ltd., The British Electric Car Co. Ltd., G. F. Milnes & Co. Ltd. to the credit of the United Electric Car Company Ltd., the writer having been appointed voluntary liquidator to each of the companies.

A NEW INDUSTRY IN PRESTON

The English Electric Manufacturing Company Ltd., of Strand Road, Preston:

In 1899 a new works was being constructed on the opposite side of Strand-road, to the works belonging to the Electric Railway and Tramway Carriage Works, Limited, on a large derelict plot of land adjacent to Preston Docks.

Dick, Kerr & Co. had taken the lead in forming a syndicate to raise the necessary capital of £200,000 for the building and equipping of a new works to manufacture dynamos, motors and controllers for electric tramcars. The members of the syndicate (otherwise known as the Equipment Syndicate) were Dick, Kerr & Company Ltd.; John Kerr, George Flett, Claud T. Cayley, Benjamin Sykes, T. S. Turnbull, Richard Prestwich, John Prestwich, George Readman, W. J. Kirk and Sydney H. Short. All except Sydney H. Short were well known in British business and financial circles, although Mr. Short did have an international reputation as an electrical engineer and designer of electric railway apparatus through his connection with the Brush Electric Company, the Short Electric Railway Company, and the Walker Company, and through the many papers which he had contributed to scientific bodies on electrical engineering matters. Sydney Short was to be the technical director with responsibility for the design of the electrical equipment to be manufactured by the new company.

The site for the new factory was a tract of 11 acres of ground, bounded by Strand Road to the east, and Preston Docks to the west, the buildings would occupy about 4½ acres of this plot. The main machine shop would be 900 ft. x 120 ft. and connected with a foundry 400 ft. x 80 ft. in area, together with a pattern shop, offices, store rooms etc; the buildings were constructed with a steel framework and brick facings and were intended to be in full operation by January 1st 1900. Electric power was to be used throughout for running of all machinery; the production capacity of the works was expected to be about 1200 equipments per annum of dynamos, motors, controllers, generators etc.

Sydney Short engaged his colleague S. B. Fortenbaugh to join him from America, as technical assistant, and the syndicate recruited most of the principal supervisory staff also from America.

On 13th April 1899, Messrs. Flett, Cayley, Turnbull and Short arrived in New York, via the SS Teutonic, in order to recruit the necessary staff and to purchase the machinery and structural materials for the new works. Contracts for at least $120,000 worth of machine tools were expected to be placed in America, and orders had already been given to E. W. Bliss & Company for the presses (Claud Thornton Cayley was a director of E. W. Bliss & Co.), and to the Kemp-Smith Company for milling machines, with remaining orders to be placed before 6th May when the entire party sailed from New York for Liverpool. Orders for the engines were placed in England with Musgraves, of Bolton.

The SS Teutonic would have been an interesting ship on which to travel to New York; she was sister ship to the Majestic, both keels having been laid in 1887. She was built by Harland and Wolff for the White Star Line under the Auxiliary Armed Cruiser Agreement, with the intention of eventual conversion to passenger service, and was the first White Star ship not to have square rigged sails. Teutonic was launched on 19th January 1889 and completed on 25th July in time to participate in the Spithead Naval Review on 1st August, and was Britain's first armed merchant cruiser, sporting eight 4,7 inch guns, and had a displacement of 9,984 gross tons. The guns were removed following the Review and on 7th August she left on her maiden voyage to New York. In 1891 Teutonic took the Blue Riband, from her sister ship Majestic, with an average crossing speed of 20,25 knots; she later improved on this with an average crossing speed of 20,5 knots. The Teutonic and Majestic have been described as the first modern liners. Teutonic had accommodation for 300 first class, 190 second class and 1000 third class passengers. She was used as a troop transport for the Boer War in 1900 and returned to passenger service between the Boer War and the Great War. She had a refit in 1916 and was fitted with 6 inch guns, participating in several convoy escorts and for troop transport in 1918, being sold and scrapped in 1921. In 1911 she was replaced in the White Star line-up by the new Olympic, sister ship to the Titanic.

The Equipment Syndicate of Manchester were prepared to begin the manufacture of electrical equipment as soon as the works were completed; manpower had been engaged and, working drawings prepared and contracts entered into for the supply of materials. The Equipment Syndicate had also entered into a number of important agreements including the engineering firm, Dick, Kerr and Co. Ltd., Kilmarnock. The Equipment Syndicate, as an entity, did not retain their interest for long, for in November 1899 Mr. Stephen Sellon, an electrical engineer of London, and Mr. William Wilson, a surveyor and valuer of Manchester, inspected the site and works on behalf of a new company. This was a newly formed company, the English Electric Manufacturing Co., whose directors awaited their report with some urgency. On 20th November 1899 the report was delivered. "We have examined the property and works now in the course of erection on the west side of Strand-road, Preston ---- we are of the opinion that the property to be acquired by the company is worth £306,805". The English Electric Manufacturing Company Ltd. bought it from the Manchester firm for £385,000. The agreement included machinery and electrical plant which the Equipment Syndicate undertook to furnish, and the board of the new owners were prepared to begin manufacture in April 1900.

The Times for 5th December 1899 reports the incorporation of this company, with a public issue of shares, the Prospectus states –

- The Company has been formed for the purpose of manufacturing every description of electrical machinery, and in particular apparatus for the equipment of electric railways and tramways. With this object in view the company has arranged to acquire works now in the course of construction and rapidly approaching completion in Preston, Lancashire, belonging to the Equipment Syndicate, Limited, of Manchester, with machinery and tools necessary to enable it to manufacture electrical machinery on a large scale, and upon the most modern and economic principles. The works, it is expected, will be ready to commence operations in or before April 1900.
- The present is a most opportune time for establishing works of this character, as the rapidly increasing demand for equipments and generators has hitherto had to be mainly supplied from American sources.
- The company will acquire the benefit of an agreement dated 17th May 1899, between Sydney Howe Short, formerly of the United States of America, and the Equipment Syndicate, Limited, whereby his valuable patent rights for Motors, Controllers, Generators and other Electrical Apparatus for the United Kingdom, India and British Colonies and Possessions, except Canada, with similar rights to all improvements he may make for a period of ten years from the date of the above mentioned agreement will be secured to this company. Mr Short (who is well known as a pioneer in the development of electrical machinery in the United States, and whose motors and Controllers are in use on Tramways in Liverpool, West Hartlepool, the Potteries and various Foreign Countries) has also consented to take a seat on the board.
- The site of the works containing 10½ acres of land is held on lease granted by the Corporation of Preston on most favourable conditions for the residue of a term of 999 years from the first day of May 1899, is centrally situated in West Strand, Preston, and connected with the London and North Western Railway, the Lancashire and Yorkshire Railway and the Preston Corporation Docks. A feature in the site is its proximity to the Electric Railway and Tramway Carriage Works, with which it is connected by railway sidings.
- The company will also acquire the benefit of certain orders for electrical equipment given by Dick, Kerr, and Company, Limited; and booked by the Vendors, which will be executed at the works of the Company. On these alone the Directors are satisfied that a profit of at least £40,000 will be earned.
- The Equipment Syndicate, Limited (who are promoters of and Vendors to the Company), have entered into a contract for the sale of the property and Works.

The whole of the Ordinary Shares were issued to the Vendors in part payment of the purchase money. None of the Preference Shares were underwritten.

Trustees for the Debenture Stock Holders were: J. E. Prestwich, Farnworth, near Bolton, and F. A. Roberts, Cornbrook Chemical Works, Manchester.

The Directors of the Company are listed as:

- Richard Henry Prestwich (Director, Blackpool and Fleetwood Tramroad Co.), Manchester, Chairman.
- George Flett, 110, Cannon-street, London, EC.
- John Kerr, 94, George-street, Edinburgh.
- Thomas Dewhurst Lingard, Booth-street, Piccadilly, Manchester.
- George Richardson (Chairman, North Metropolitan Tramways), London.
- Sydney H. Short, 110, Cannon-street, London, EC., Technical Director.

Solicitors for the Company: Sutton, Elliott, Turnbull, and Mayne, 13, Spring-gardens, Manchester.

Secretary (pro tem.) and Registered Office: Frederick Crimes, 13 Spring-gardens, Manchester.

Frederick Crimes, a Solicitors Clerk, is found representing a number of the companies, in the same role and on behalf of Sutton, Elliott, Turnbull, and Mayne, with which Dick, Kerr and Co. have an interest, including: The Electric Railway and Tramway Carriage Works, Ltd, the Blackpool and Fleetwood Tramroad and the City of Carlisle Tramways.

A contemporary American source provided the following assessment of the English Electric Manufacturing Company and its promoters; quote –

> The well known energy and indefatigable determination of Dick, Kerr & Company in pursuit of electric railway business in Europe have won them hosts of friends, and there is no question about their ability to fill their new electric works to overflowing with orders. One of the interesting features of the entire transaction, and one characteristic of British quickness to take advantage of opportunities for investment is found in the fact that the entire capital for the construction of the works, £200,000, was subscribed in about 30 minutes, and no subscription of less than £10,000 was accepted, while the desire to obtain larger amounts than could be allotted was urgent.
>
> One of the unpleasant features about this new enterprise is that Mr. Short's friends in America will lose him for several years at least, as in pursuance of his contracts with the new company he will have to take up residence in England. He returns with Mrs. Short to England on May 6, and will shortly be followed by the remainder of his family. He has leased for several years a beautiful house and estate in Weybridge, one of the most charming of London's suburbs, and has been fortunate in securing one equipped with all the modern improvements, including a system of heating by hot water, a rare thing to find in England.

Start of Building. English Electric Manufacturing Co. West Works July 1899.

English Electric Manufacturing Company Limited. Construction of the Foundry in February 1900.

The new West Works Boiler House c1900. English Electric Manufacturing Co. Ltd, Strand Road West Works.

English Electric Manufacturing Company Limited c1900. Showing the installation of the three Boilers. Manufactured by J. Foster and Sons, of Preston.

Power House Engine "Irene". Manufactured by, John Musgrave & Sons of Bolton.

English Electric Manufacturing Company Limited c1900. Machine Shop.

Manufacture starts English Electric Manufacturing Co. Ltd, Strand Road, West Works c1900.

English Electric Manufacturing Company Limited c1900. Early picture of the Works Interior, not yet fully occupied.

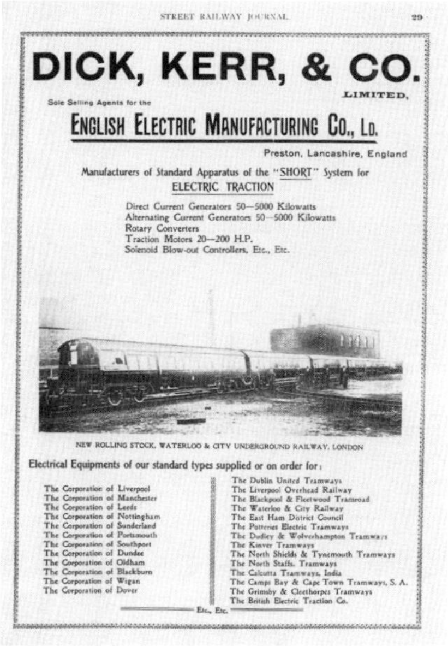

Dick, Kerr and Company Limited.
Advertisement from 1899.

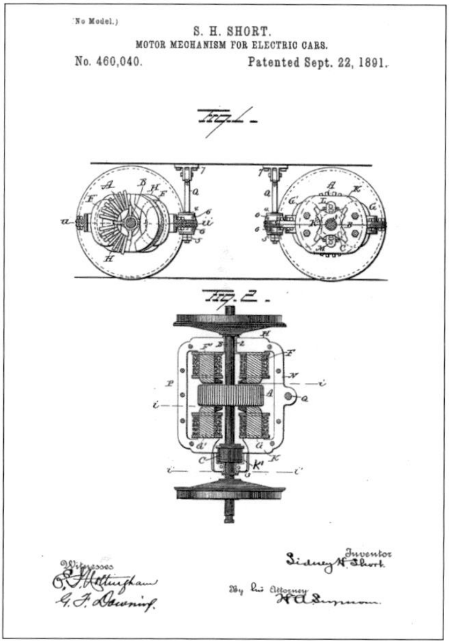

Patent for a Motor Mechanism
Sydney Howe Short September 1891.

Dick, Kerr and Company Ltd.
Advertisement from 1902.

Dick, Kerr and Company Ltd.
Advertisement from 1902.

English Electric Manufacturing Company Limited c1900 Pattern Shop.

English Electric Manufacturing Company Limited c1900. Workers at their Benches manufacturing Detail Parts.

On 2nd December 1899 the Lancashire Evening Post under the headline GREAT ELECTRICAL ENTERPRISE AT PRESTON gave a very enthusiastic report on the new undertaking –

> The great development of electric tramways and railways is one of the principal phenomena of the rapidly-moving times we live in, and England is quickly putting herself on a level with foreign countries in the extensive use of these methods of traction and locomotion.---The English Electric Manufacturing Company, whose prospectus has just been issued, is the first big step made in the direction of manufacturing electrical tramway and power transmission machinery.--- Had the works been in operation a year ago it would have been unnecessary to send to America for the electric locomotives and large generators to be used in the operation of the Central London Underground Railway.--- Orders have already been booked from a number of English cities to employ the works for nearly a year on tramway motors, generators and general electrical equipment; and it may be interesting to note that orders for 400 tramway equipments have been received from the Continent, and that the company's engineers are at work on a new line of electrical mining machines for use in South Africa. The prospects are certainly bright for this class of industry, and Preston, which it is gratifying to note is enjoying an amount of trade prosperity which has been the subject of many public congratulations, will benefit still further by such enterprise.

On 22nd June 1900 the Engineer journal had a detailed description of a 'New Electric Manufacturing Works'–

> The works of the New English Manufacturing Company, which were formally opened on Thursday, June 14th, are situated in the Strand-road, Preston, Lancashire. Representatives from The Engineer were invited by the directors to make an inspection of the works, 'and as the ordinary practice of this country has in so many cases been departed from, we feel sure that a description of them will be of interest to our readers'. This was seen from the outset as an experiment of bodily planting an American shop in England, arranging and equipping it on American lines, and working it to the American system with British workmen. There was reported to be a small nucleus of American workmen- some ten or eleven-nearly, if not all of them foremen, but the 'workmen proper' were entirely English, drawn from the surrounding district. Of the result of the system of management, and of the way the British workmen will behave when working on lines to which he has not been accustomed, it was considered 'too early to speak'.

The report expressed some concern regarding circumstances in this country which do not obtain on 'the other side' and what trade unions would have to say to American methods.

There appeared, however, to be no apparent reason why the experiment should not be a success, and the men seemed to be on excellent terms with their employers.

The length of the working week was fifty hours; the works opening at 7.30 a.m. and closing at 5.30 p.m. There was an hour for dinner between 12 and 1, but there was no breakfast interval. On Saturday work commenced as on other days at 7.30 a.m. and ended at 12.30, when the men were paid. Overtime, which could only be earned after a full 50 hours had been worked, was paid for the first two hours at the rate of time and a-quarter, and for any time afterwards at time and a-half. There were some 450 men employed, but inasmuch as the works were not yet in full swing, the number was expected to be augmented.

The power house which stood alone contained three boilers, each 32ft long, and 8ft 6in diameter, of the Lancashire type, and made by J. Foster and Sons, of Preston. The chimney was described as circular in form and 200ft high and like the rest of the building, built of red brick. There were two generating sets fixed, one of which was at work continuously driving all the machinery and doing the whole lighting of the factory. The engines, manufactured by John Musgrave and Sons, Limited of Bolton were, with the boilers, two sets of Green's economisers and the metal ladles for the foundry the only machinery of English manufacture to be found in the whole works, machinery and equipment having been sourced from all parts of the United States. Almost all the machinery was driven from overhead shafting supported from longitudinal beams, the works had about 250 individual machine tools all of the most recent type, practically every firm in the United States being represented. The two largest machines were a 24 ft, boring mill made by the Niles Tool Works, and a universal facing, boring and drilling machine by the Betts Machine Company, of Wilmington, which were stated to be 'both beautiful bits of work'.

The electrical supply to the manufacturing shop was fed through a tunnel and carried on porcelain insulators. The supply was then taken to a large slate switchboard for distribution to the various machines, lights, &c. The lighting was by fifteen Jandus arc lamps along one side wall and eighteen on the other. In addition there were incandescent lamps over every machine-sometimes two.

The blacksmiths shop was found in the machine shop with four Buffalo Forge Company forges, the blast for which came from a small Buffalo blower.

This was followed up on 20th September 1901 with a performance report on a 400 kilowatt direct current generator, which had recently been manufactured by the company; the introduction to the report is of particular interest. –

> The company has since it started work in June last year --- completed and shipped nearly 1500 tramway motors of sizes varying from 25 to 100 horsepower, together with the necessary equipment for some 750 cars. But it has not only been engaged on tramway motor work. It has also been manufacturing generators of considerable size, and it despatched from the works about four months ago the first of these.

'English Electric and its People', the company newsletter up to the mid 1960s included many reports on the retirement of long serving employees. One of these, who started work with the company in the early days with the English Electric Manufacturing Co., and retired from the post of Publicity Manager, referred to the many outstanding men he had met in his work with the company, among whom were George Flett, Chairman of Dick, Kerr, Walter Rutherford, Managing Director, Claud T. Cayley, Sir Charles Ellis, John Kerr-at one time Member of Parliament for Preston, Bernard Firth and Sir John Aspinall (General Manager, Lancashire and Yorkshire Railway). Another ex-employee who was interviewed by the journal in January 1966, at the age of 85, joined the English Electric Manufacturing Co. in 1900 during the building, construction and installation of the production plant. He noted that "this company was closely associated with the Dick, Kerr Company of Kilmarnock". He names the original team of designers as Prof. Short, the Von Zweigbergk brothers and Prof. Zani. He had the good fortune to assist Prof. Short with the running and testing of the first dc generators on order for St. Helens Corporation. He also assisted Profs. Fortenbaugh and Ryan with the first traction motor tests to be carried out at the Preston works. The first traction motors were delivered to Liverpool and Manchester tramways.

On 21st March 1900 the annual summary of persons holding shares included:

			Ord.	Pref.
Richard Henry Prestwich	13 Marsden Square, Manchester	Cotton Spinner	200	1
Benjamin Sykes	33 Winckley Square, Preston			1
Thomas Dewhurst Lingard	10 Booth St., Piccadilly, Manchester	Solicitor	200	1
Sydney Howe Short	110 Cannon Street, London	Engineer	200	1
Thomas Strover Turnbull	13 Spring Gardens, Manchester	Solicitor		1
George Flett	110 Cannon Street, London	Contractor	200	1
Francis Owen Mott	110 Cannon Street, London	Clerk		1
John Kerr	94 George Street, Edinburgh	Engineer	200	-
George Richardson			200	-
Richard Henry Prestwich	13 Marsden Square, Manchester	Cotton Spinner	)	
Thomas Strover Turnbull	13 Spring Gardens, Manchester	Solicitor	) 31,800	
George Sillivan	14 Albert Road, Southport		)	

On 25th March 1902 Richard Henry Prestwich and Thomas Strover Turnbull are shown to be holding 38,800 ordinary shares. They were partners in the Equipment Syndicate, along with Dick, Kerr & Company Ltd.; John Kerr, George Flett, Claud T. Cayley, Benjamin Sykes, John Prestwich, George Readman and W. J. Kirk, the principal investors in the construction of the works and which before its completion, was taken over by Dick, Kerr and Co., the Syndicate continued to be the largest single shareholder. The works operated as the English Electric Manufacturing Company, although managed from the Dick, Kerr head office in London where some of the Design Staff were located and headed by Professor S. H. Short; George Flett was Managing Director of Dick, Kerr and Co. Ltd. In 1901 James Conner was recruited from his position as Works Manager at the Britannia works of Dick, Kerr at Kilmarnock to be Works Manager in Preston.

The Agenda Book kept by S. H. Short for the period 24th April to 25th September 1902, records the weekly payment of interim cheques included in his briefings to the board. The agenda for the meeting to be held on 24th April 1902 at the Company's Offices records:

Payment of interim cheques total £4,796-0-10 including -	
J. Conner	£83-6-8
T. Von. Zweigbergk	£50-0-0
G. Von. Zweigbergk	£24-0-0
Payment of interim cheques total £20,282-12-3 including -	
R. H. Prestwich	£130-5-0
G. Flett	£120-0-0
S. H. Short	£120-0-0
G. Richardson	£125-0-0
T. D. Lingard	£104-4-0
J. Kerr	£120-0-0
Equipment Syndicate Co.	£19,506-1-7 (*August payment* £13,237-8-3)
F. A. Roberts	£27-10-10
J. C. Prestwich	£23-10-10
Company assets	£42,129-4-2
Liabilities	£18,247-13-0

It would appear that the Syndicate are being paid in instalments.

Messrs Richardson and Lingard are also Directors of the City of Carlisle Electric Tramway Company.

For the period 1900 – 1905 the Management Committee minute book, compiled by George Flett, provides an interesting chronological record of the activities of the Company in its early years.

- On 16th May 1900, office hours and workmen's hours are discussed, a dining room for staff to include cooking and heating and portable office accommodation for the foremen are among the topics under discussion. Payment to the Zweigbergk brothers and Fortenbaugh is considered and John Barr is appointed General Clerk on a salary of £110-0-0 pa.
- On 26th June, A Tool Designer is to be recruited with arrangements left to Mr. Short. Piecework in the workshops was resolved to be established on a basis of about 15% less than Mr Short paid in America.
- On 17th July the salary of T. Von. Zweigbergk was increased to £50-0-0 per month.
- On 6th September, cottages owned by the company were not being let as quickly as planned and it was agreed they should be let to employees of the Carriage Works. A later report approved the letting of 57 cottages.
- In October there is a clear indication that finances are being stretched when loans to the total of £26,500-0-0 were agreed to be renewed and the Secretary was requested to curtail purchases as much as possible.

Invoices outstanding included Messrs Dick, Kerr & Co. for 50 equipments totalling £11250-0-0.

A letter had been received from Dick, Kerr & Co. proposing a joint exhibit at the Glasgow Exhibition to include the Carriage Works with expenses being shared.

It is quite clear that, at this time, the three companies, Dick, Kerr of London and Kilmarnock, the English Electric Manufacturing Co. and the Electric Railway and Tramway Carriage Works in Preston operate as separate entities albeit having some common management representation. Dick, Kerr being the parent company with the others having independent financial and business structures. The Carriage works, later the United Electric Car Co. Ltd., is perhaps the more independent of the two with the original financial backers from Manchester, having the largest single shareholding, amongst its Directors. Nevertheless, the three companies clearly have a common bond, if only by the presence of John Kerr on their respective Boards, but become increasingly interdependent with regard to their business interests, customers and range of products.

- On 15th November 1900, Sydney Short applies for Royalties on equipments sold at 30th September. The Secretary was instructed to inform Mr Short that no complete equipments were invoiced on 30th September.

On the same theme, on 20th July 1901 the Company served preliminary notification on the Solicitors of the Central London Railway regarding infringement of patents assigned to the English Electric Manufacturing Co. by Mr Short.

Sydney H. Short had brought his many American patents with him when he joined the Company and was allowing the Company to make use of his inventions, although it would appear from the above that he had not surrendered his patent rights.

- On 11th January 1901, a decision was taken to discontinue the present management committee and that the General Manager (*James Conner*) should take over the management of the business.

Although still living in Kilmarnock this probably marks the approximate date that James Conner was appointed to the position of Works Manager at Preston.

- On 24th January it was agreed to pay the travelling expenses of London Staff moving to Preston. The London staff arrived in Preston on 4th March 1901.

The office space vacated, at 110 Cannon Street, by this move was subsequently taken by the J. G. Brill Co. of Philadelphia, this address is given on the company letterhead as their London office. J. G. Brill was the main supplier of trucks to the Electric Railway and Tramway Carriage Works and later the United Electric Car Co.

J. G. Brill had opened a London office at the beginning of 1899, at 17 Victoria Street, at a time when Brill was seeking to expand into Great Britain, and moved to 110 Cannon Street in 1901. This office served as headquarters for their European Manager, Maskell Curwen, as he worked towards establishing a European affiliation.

In 1901 Brill took a serious look at establishing a plant overseas, in England. By this date car builders outside the United States had become proficient in the business of tramcar manufacture and high tariffs discouraged the import of cars from the United States. The situation with trucks was a different matter and Brill trucks were in great demand all over the world. During 1900 nearly 2,800 trucks were sold by Brill to railways outside the United States, with 1,700 sold to Great Britain alone.

This was the situation when the Brill Company decided to send John Brill to England to look at the possibility of establishing a suitable location for a truck building plant. John viewed potential sites near Birmingham, London and Preston, as the company had definitely decided to build a plant somewhere in England. Martin Brill, as Company President, visited England, in November 1901, to complete negotiations, but at the last minute the deal fell through for reasons unknown.

What ultimately transpired with Brill truck sales to Great Britain was that the majority were purchased through Dick, Kerr & Company and the Electric Railway & Tramway Carriage Works, Ltd., of Preston. These were fitted to trams manufactured for both the United Kingdom and export markets.

The J. G. Brill Company eventually established a subsidiary company in France, Compagnie J. G. Brill at 14 Place de Laborde, Paris. This facility commenced operations in June 1908.

English Electric Manufacturing Company shipments during January 1901, to the value of £14,141-0-0, included- motors, controllers, rheostats, resistors, fuse boxes and circuit breakers. Orders procured amounted to £27,383-8-4 and wages paid to £3,610-13-1.

Orders received during the first half of 1901 appear to have been extremely variable from a low of Nil on 16th February to a high of £21,442-10-10 on 22nd June, whilst shipments look reasonably healthy, although mostly fulfilling existing orders, from a low of £2,734 on 9th February to a high of £8,553-0-0 on 11th March.

- On 2nd February 1901 the works closed for the funeral of Queen Victoria.
- On 15th February, a shunting locomotive was ordered from Peckett and Sons of Bristol at a cost of £920-0-0.
- At about 7pm on 12th March 1901 a fire broke out under the floor of the drawing office, above the boardroom apparently caused by defective electric light wiring and leakage of gas.
- In July there was concern that orders were 'not being received to the value that might be expected'.
- On 14th December, it was reported that 'there is slack in both motor and generator departments' a large number of orders had been received for generators but these were still in the design stage, more orders were in prospect.
- During the period February to April 1902 orders for motors and generators were received from – Barrow, Ilford, Yates & Thom (Blackburn), Govan, Rothesay, Cardiff, Southampton, East Ham, Portsmouth, Sheffield, York, Ashton Manor, Bristol, Hong Kong, LCC and Liverpool. It was noted that "English Electric Manufacturing's capability for motor production far exceeds that of the carriage works production capability for cars."

Electric Generators for Bolton Corporation:

On Thursday 16th January 1902 representatives of the Engineer Journal accepted the invitation of Dick, Kerr, and Co. to join a small party specially invited to visit the works of the English Electric Manufacturing Company, Limited, of Preston, Lancashire, in order to be present at the test of two 1,100 kW generators which have been built by the company for the Bolton Corporation. These machines were to be used for the supply of energy both for lighting and tramway work.

The Merger

The English Electric Manufacturing Co., Ltd. went into voluntary liquidation on 31st March 1903 the capital being acquired by Dick, Kerr, and Co. Ltd. The final meeting return was registered on 5th March 1904. This move was regarded at the time as 'enormously strengthening the industrial position of Dick, Kerr & Co., who are rapidly establishing themselves in the front rank among British manufacturers'; Dick, Kerr had 3,850 employees, 'putting them at 77th place in the list of largest employers'. It was accompanied by a significant change in the structure of what had become a diversified, geographically-separated organisation. The London offices, at 110 Cannon Street, London, was the headquarters where the Board and the Management and Sales Committee sat, while each factory, in Preston and Kilmarnock had its own Works Committee, with a representative from the Board, usually George Flett, acting as the medium of communication between the senior executives and the manufacturing plants.

A newspaper report dated Tuesday 22nd September 1903 gives an account of a British Electrical Consolidation –

A Manchester correspondent states that negotiations are in train for some important negotiations or consolidations in the British Electric construction and car-making industry. Considerable reticence is displayed as to the names of the companies or firms interested; but it is understood that Messrs. Dick, Kerr, and Company, Limited of Glasgow, Preston, and other places, who recently took over the English Electric Manufacturing Company, Limited and are intimately connected with the Electric Railway and Tramway Carriage Works, Limited, of Preston, are at the head of the movement.

This was the first report alluding to the takeover of G. F. Milnes & Co. and the British Electric Car Co. Ltd., by the Electric Railway and Tramway Carriage Works to form the United Electric Car Co. Ltd., which was brokered by Dick, Kerr & Co. George Flett accompanied by E. A. Stanley, Works Manager at Preston, had visited the Hadley Works of G. F. Milnes on 3rd – 4th September 1903 to examine the books. The executive management of the United Electric Car Co., Ltd., however, remained independent of Dick, Kerr & Co., until the formal takeover in 1917.

By 1903 Dick, Kerr had become the leading firm in the British electric traction industry and, with the United Electric Car Co., supplied more trams and equipment than any other manufacturer. In taking control of the English Electric Manufacturing Company, Dick, Kerr had gained a significant advantage over its competitors; they now had the capability to deliver the complete package for a new tramway system, power generation, civil engineering works and tramcar manufacture. In the early years contract submissions and design activities would be centred on the London office, contracts for power house and tramway construction work would be allocated to Kilmarnock, with the manufacture of power generation equipment, tram motors and controller's etc. being carried out at the works of the English Electric Manufacturing Company on Strand Road, Preston, although engines to drive power generators and alternators would be sub-contracted to other manufacturers such as Willans & Robinson. If the contract included the supply of tramcars this would be sub-contracted to their associate company the United Electric Car Co. Ltd, on Strand Road, Preston.

The By-Election

During the run up to the Preston by-election, in 1903, the Lancashire Daily Post gave a thorough description of the business of Dick, Kerr & Co. which is worth quoting from in part since it paints a vivid picture of the company and its business philosophy –

> The candidature of Mr Kerr for Preston naturally brings to the forefront the great industrial enterprise which he and his partners have founded in our midst under the title of Dick, Kerr & Co. A few years ago the Strand-road with its great derelict works of the old North of England Iron Company was a silent avenue of traffic, with a great expanse of the marsh as a playground for the boys and girls residing on that side of the town. During the last few years an enormous transformation has been effected, and the industrial life of Preston has strengthened in a degree which must very materially add to its potential value and prosperity. The works are the outcome of Mr. Kerr's resolve to show that we in England could produce electrical equipment quite as good, if not superior, to the great works which have been founded in the United States and in Germany, where the absence of legislative enactments promoted the speedier development of public electrical systems. One need not go into the whole history of Mr. Kerr's earlier association with the great Kilmarnock manufacturing industries, suffice it to say that at the Kilmarnock works general engineering, locomotives, steam engines and finished ironwork of these descriptions are made by the firm which there employs 500 men.
>
> ---The business at Preston was on absolutely new lines. The firm had for some time contracted for the laying down and equipment of tram systems, and they became, with that business-like discernment of the needs of the hour, the pioneers of that electrical system which has revolutionised the light traffic of our towns and cities and interlinked the busy communities and villages in every county. The orders for cars used to have to go to America and Mr. Kerr and his colleagues in partnership naturally asked themselves why we in England should be behind hand in turning out speedily artistic and scientifically constructed trams, and why we should not further be able to supply fully, absolutely, and smartly the most modern electrical equipment, which the inventive genius of the day has fashioned, into practical forms. America certainly had an early advantage. She rapidly went ahead with electric tramways, and manufacturers in the United States were thus enabled to standardise traction machinery. Messrs. Dick, Kerr and Co., have shown that what could be done in America could be repeated here, and so Preston having been selected as the venue for the great enterprise, they started the company under the title of the English Electric Manufacturing Co., Ltd., in which their financial interest was supreme.

They had previously purchased and equipped the old Wagon Works as a great car making depot and turned out their work with phenomenal rapidity, employing large numbers of skilled artisans. The two firms were practically the same, only for a time under different titles; it being found inconvenient for business that there should be two distinctive titles to the firm, the style of both firms has been made uniform, and will now and henceforth be known as Dick, Kerr and Co., Ltd. For the moment we need not enter into any lengthy description of the old Wagon Works, where trams are being constructed with a rapidity which the contracts of the company have demanded, and which has often involved working day and night.

---The successful competition of an English firm with foreign manufacturers has not been confined to this country, machinery having been constructed for foreign tramways, as well as for India and the Colonies. The most recent orders include important equipments for Durban, Calcutta, Mandalay, Singapore, Hong Kong, Sydney, Alexandria, Canada, Ceylon, Colombo and Japan.

---The demand for electric tramway machinery in this country has been enormous, and one has to thank the enterprise of Dick, Kerr and Co. for preventing to a great extent the diversion of orders which but for their works would have gone to foreign firms. There is thus established here an object lesson which ought not to be lost on the industrial captains of England. Too much fuss and nonsense have been made as to the superiority of foreign competitors, and if the rest of the leaders of British enterprise would only take a leaf out of the book of Dick, Kerr, and Co. they would very soon lower the flag of fierce rivals in other progressive nations.

--- Of the 12 acres of land comprehended about 5 acres are under roof Alterations are at present proceeding for additional workshop accommodation at the equipment works. The buildings, in accordance with the best practice of the day, are one storey high throughout except the offices. They are laid out on the line of the most recent American engineering works.

---The machine shop is divided into two enormous bays, one on either side of a centre stanchion. One of these bays is devoted entirely to the manufacture of traction motors and details of equipment, while the other is used for the production of heavier machines, such as large dynamos, alternators, rotary transformers &c.

---The largest tool in the works is a boring mill swinging 24ft for boring out armature heads to receive the engine shafts, to finish the armature end rings for large generators and to make the large commutators for these machines. The mill weighs over 100 tons.

---Within the grounds there has been erected a directors clubhouse. The arrangements for the comfort and entertainment of the workmen are considerable. Alongside the foundry it is arranged that extensive bathrooms shall be erected, provided with hot and cold water plunges &c, in which each of the men employed in the foundry will have his own special locker for his private use. A portion of the company's funds is being devoted to the erection of a large group of comfortable workmen's dwelling houses upon an excellent site close at hand to the works. It is also intended to erect a large clubhouse with gymnasium and reading rooms, &c. Mr Kerr and his colleagues are taking an active part in promoting technical education, knowing the importance which the young engineer, with a fair technical education and a sound practical training, has played in the building up of the industry in connection with the great German and American companies. They are arranging for the delivery of a regular course of lectures on electrical engineering throughout the year to the young men, and have generously offered to assist in the equipment of the electrical school at the Harris Institute. Mr Kerr some time ago offered to pay a sum of £500, on condition that a certain larger sum was realised, and this offer is still open.

- On 22nd July 1904, Mr. Flett proposed that the Preston Works Committee be set up along similar lines to the London Management & Sales Committee, the first members to be: Messrs Conner, Zani, Zweigbergk, Dodd, Buckmaster and Barr. The chairman to be Mr Flett or Mr Rutherford then failing these Mr Conner. *George Flett and Walter Rutherford were both from the London Office.*

In his report Mr Conner gives production output as currently £7,000 to £8,000 weekly, need to aim for £8,000. 1,100 hands were employed, with office and outside staff total 1,400 including permanent night staff (shift) of about 80.

Contracts nearing completion and 'other customers' were- Gloucester, Worcester, Preston, Plymouth, Reading, Brighton, Northampton, Alexandria, Delagoa Bay, Leeds, Wakefield, Swindon, Tokyo (60 cars and equipment), Lancashire & Yorkshire Railway, Sydney, Portsmouth Dockyard, Barrow in Furness and the War Office.

- On 28th July, Mr Flett proposed that the London Office should forward weekly reports from Manchester & Glasgow (Kilmarnock) and any from local Agents. Further customers included – Bath, Leicester, Nottingham, LCC, Cardiff, Wakefield and Liverpool Overhead.

- On 5th August 1904, Mr Rutherford in a 'cash statement' reported for the year ending June 1904: £281,000-0-0 from Head Office
  £272,000-0-0 goods delivered

Further customer enquiries included – Leeds, Bristol, Durban, Perth, Blackburn, Woolwich, Bournemouth (Gas & Water Co), Manchester Ship Canal, Tynemouth, Potchelstroon Military Camp, Bath, Sheerness Dockyard and Colne.

Equipments required included- 1,000 kW turbo generators (ac), 250 kW (dc) rope pulley, switches, capstan motors, stationary motors, dynamos, and generators of various sizes.

Dick, Kerr Works Strand Road c1910. United Electric Car Company on the right, east works. English Electric Manufacturing Company on the left, west works.

January 1904, Dick, Kerr & Co., Ltd. after the annual lunch for the production staff. Back row, left to right: Messrs. Wilson, R. G. Devey, Bateson, Kay, G. Pownall, J. Barr (Company Secretary), Whitehead, M. A. Cox, Barnes, Davidson. Front row, left to right: Messrs. Green, Elliott, J. Connor (General Works Manager), G. Flett (Managing Director), W. Rutherford (General Manager), Salisbury.

The Dick Kerr Works on Strand Road, Preston. Group of Managers posing beside a large Armature Casting.

There is a gap in the records between 1905 and 1908.

- On 15th February 1908, the Works Committee comprises – Messrs Conner, Zweigbergk, Balcombe, Jacklin, Facer, Green and Barr; Chairman Mr. Conner.

Mr Zweigbergk proposed increasing the range of crane motors by two more sizes, 2hp and 15hp, to complete the range and enable the Company to compete more effectively. This was to be put to the London Committee. An extension to the power house was proposed to provide power for driving and testing the large alternators for the LCC, presently in manufacture.

- On 28th February, Mr Flett reported on the problem of high prices being quoted and consequent loss of orders to competitors. He gave examples based on prices submitted to the Aluminium Co. for stationary motors-

Our price	£2,544
Edison Co.	£1,633
Electric Ordnance	£2,633
General Electric	£1,729
Johnson & Philips	£1,862
McAlpine	£1,879
Siemens	£1,925

Competitor's machines were considered to be of inferior quality, in the case of the LCC machines it was understood that the Westinghouse machines were inferior, Mr Zani had pointed out that "they were not up to specification quality having been assessed by weight comparison" (the amount of copper used). Mr Flett concluded, "In his opinion business should not be fought for on a question of quality rather than on lower price with reduced quality".

On 21st January 1909, new contracts included – Loch Leven (machines), Vancouver (locos), St. Pancras (machines) and Portsmouth (turbine).

East Ham Power Station:

Parliamentary sanction for the introduction of the first municipal electric tramway in the metropolitan district of London was obtained by the District Council at East Ham in 1898. At the same time the Council also gained approval for an electric lighting system. Tramway privileges were extended in 1900, so as to cover a total of 2¼ miles of double track and 6 miles of single track, of which 4½ miles had been completed by September 1901. The system was particularly interesting at the time, not only from the fact of its being the first municipal electric line in London, but also because it combined a lighting and tramway service. The contracts were awarded in September, 1899, and Dick, Kerr & Company was the successful bidder for the complete installation.

The tramway buildings included the power station, measuring 177 ft. by 42 ft., contained three 225 kW generators, each direct coupled to a 350 hp engine and one 100 kW generator coupled to a 150 hp engine; the engines were built by Musgrave & Sons of Bolton.

The generators, which were direct driven, were manufactured by the English Electric Manufacturing Company Ltd. of Preston, and each could be used for either lighting or traction work. When used for traction purposes they worked with a compound winding and had a variation of from 500 volts to 550 volts. When employed for lighting they ran as shunt-wound generators, having a full load capacity with a maximum of 580 volts at full speed. The main switchboard consisted of thirteen panels of white marble mounted on a raised platform, and was made by the English Electric Manufacturing Company Ltd.

The overhead line consisted entirely of side-pole and bracket construction and the design of the pole bases were of the type adopted by Dick, Kerr & Company who was the contractor for this work.

The rolling stock initially consisted of fifteen top seat cars, known as the Liverpool type, having the 'Bellamy' reversed stairway. These cars were supplied by Dick, Kerr & Company and manufactured by the Electric Railway & Tramway Carriage Works at Preston. Each car was mounted on a Brill 21-E truck and had a seating capacity of fifty six passengers, twenty two inside and thirty four outside. Each car had two motors of the 25-A type designed to give a tractive effort of not less than 1000 lb when running at 10 miles per hour. The controllers, like the motors were of the well known 'Short' (*Sidney H. Short*) design of the solenoid blow out type with rheostat braking capable of controlling the car in an emergency.

The tramway system was put into operation on Saturday 22nd June 1901.

Turbo-Alternator for Sydney:

At the beginning of June 1907 representatives of the Engineer journal were invited by Messrs. Dick, Kerr, and Co. to visit with them the works of Messrs. Willans and Robinson, of Rugby, to inspect a large turbo-alternator destined for Sydney; the alternator having been made by the former firm and the turbine by the latter.

Willans and Robinson of Thames Ditton and Rugby was founded in 1880 by the partnership of Peter Willans and Mark Robinson and in 1916 became part of Dick, Kerr, and Co.

Braking Tests in Leeds:

Dick, Kerr & Co. Ltd also became involved, with the customer, in trials aimed at improving the performance of their products.

On 25th November 1907, J. B. Hamilton, general manager of the Leeds City Tramways, gave a public test with two cars, one No. 87 fitted with the latest type of Westinghouse magnetic brake, and the other No. 270 with a new electro-mechanical track brake designed by the tramways' own engineer. Both cars were of exactly the same weight and carried the same equipment except for the brakes. They had Dick, Kerr 35-A motors and British Thompson-Houston B-13 controllers and were mounted on Brill 6 ft wheel base trucks; they both weighed unloaded, 23,600 lb.

Experiments with Breaking High Voltages:

Reported in the Street Railway Journal for 8th February 1908 –

> Some interesting experiments have recently been conducted at the works of Dick, Kerr & Company, at Preston to determine the ability of the Von Zweigbergk metallic shield blowout for breaking high voltage direct current, and incidentally the practicability of high voltage direct current railway operation. The tests were made with a specially designed blowout coil 6½ ins. In diameter and weighing about 35 lbs. Direct current at 3,200 volts was applied with twelve 40 hp, 300 volt railway motors connected in circuit, two in parallel and six in series, in order to give an induction discharge similar to that which would be produced by two 175 hp, 1,500 volt motors in series. A current of from 75 to 100 amps was broken several hundred times but hardly any traces of wear could be detected on inspecting the contacts. The results have been so successful that equipment for 4000 volts direct current is now being built by Dick, Kerr & Co. The equipment will consist of two motors each of 1900 volts, but capable of working up to about 2,000 volts and of about 200 hp capacity each. These motors will be connected permanently in series to allow for a line voltage of about 4,000. The Von Zweigbergk Controller Company, whose headquarters are in Cleveland, Ohio, is now manufacturing blowouts for this voltage.

Turbine Plant for Cairo:

In 1908 Willans & Robinson, of Rugby supplied a turbo-alternator to the Cairo Electric Railways & Heliopolis Oasis Company to supply power for working railways in the suburbs of Cairo running out into the Heliopolis Oasis. The turbine, which was built to the specifications of A. Della Riccia, of Paris, was of the Willans-Parsons type of 3,000 kW normal capacity, running at a speed of 1,500 rpm and capable of an overload of 25 per cent for two hours and 50 per cent for short periods. The turbine was direct coupled to a three-phase alternator manufactured by Dick, Kerr & Company, of London and Preston.

New Developments

The summary of the annual report of Messrs. Dick, Kerr & Co. Limited published on 21st September 1909 created a 'good deal of interest' –

> The directors reported, with expression of regret, 'that the severe competition referred to in the previous report had continued throughout the past twelve months in a more acute form; the quantity of work carried out in all departments had been considerably below the average'. The profits earned amounted to £28,168, and out of this had to be paid Debenture interest and trustees fees, and there had been reserved the sum required to provide for the premium payable on the redemption of the present Debenture stock. A sum of £12,693 was absorbed by these items, leaving £15,474, to which must be added the profits brought forward from last year - £81,566 – making a total of £97,041 available for appropriation. Out of this the directors propose to pay a dividend of 6 per cent per annum on the Preference share capital; a dividend of 6 per cent on the ordinary share capital of £260,000; and to take on the balance of £63,141, thus making a nice slice in the carry-forward capital. In 1907-8, and in previous years, dividends of 10 per cent on ordinary shares, several times accompanied by substantial bonuses, had been paid.

Twelve months later and the Ordinary General Meeting of 29th September 1910 at the Cannon-street Hotel, Mr John Kerr presiding –

> The Chairman, in his opening remarks, referred with great regret to the death of Mr. George Flett, the late managing director, as the result of a motor accident at Birmingham on July 27th. Mr. Flett had been associated with the company for nearly 30 years. "It added to the sadness of the affair to think that certain work which their deceased colleague had strenuously toiled at and laid the foundations of might so soon materialize and that he would not see the harvest reaped". Mr. Rutherford , who had for a number of years acted so ably as assistant managing director, had been appointed managing director. Other arrangements as to the management were receiving the close attention of the Board.
>
> In his review of past performance the Chairman painted a gloomy picture of "extreme competition in the electrical trade", and "In times of depression all industrial concerns have to face restricted profits, but there is a point where discretion should cry a halt".
>
> "After three years of dull times a period seems at hand when there will in many directions be a great expansion of electrification in this country and abroad". "The future is brighter, and, without holding too optimistic views, your directors are of the opinion that the work at present on hand and in prospect makes the outlook distinctly encouraging".

The Manufacture of Steam Turbines: - "After due consideration your board decided to take up the manufacture of steam turbines, and arrangements of a satisfactory character have been concluded to make turbines under the Bergmann patents. It is hoped that this departure will be the means of increasing our business in large alternators, for which we have a good reputation".

Robert Livingstone was appointed designer with responsibility for steam turbines, the manufacture of which were under a German licence. He spent some months in Berlin at the patentee's works, collecting technical data and manufacturing particulars.

The death of George Flett must have been a serious blow to the Company and his close colleagues. Having been associated with the Company and John Kerr for more than 20 years he was arguably the driving force behind most, if not all, the significant developments concerning the growth of the Dick, Kerr business. Coming at a particularly critical time in the fortunes of the company, clearly indicated in the meeting reports, his untimely death was all the more regrettable. Nevertheless it would appear that recovery plans were already in place and work was in hand to add another, and completely new, product to the range already offered by the company.

At the meeting of shareholders on 28th September 1911, John Kerr reported that –

"Keen competition for orders had continued in all departments. The development of electrical power in ironworks and the textile trades and many other industries continued. If this country did once lag behind Germany and America in the adoption of electricity, it might fairly be said that the greatest had now been awakened. The competition for business was keener between home manufacturers, but it was increased by our markets being open for Continental and American work. They had every opportunity of maintaining their output, and to anyone who knew anything about manufacturing, the question of cost was largely one of output".

"Their efforts in the manufacture of steam turbines were meeting with satisfactory results. In view of the enormous demand for filament lamps, they had decided to add an electric lamp factory to their establishment. The prospects for the coming year were encouraging. As soon as the international difficulties were settled a large amount of foreign work that was held in suspense would be proceeded with. He hoped the Government would deal in no uncertain manner with the question of peaceful picketing".

The report for the year stated that –

Profits amounted to £37,994. After paying Debenture interest and trustees' fees, etc., there was a balance of £25,934, to which was added the profits brought forward from 1910 of £42,258, making a total of £68,192 available. A dividend of 5 per cent was paid on the Ordinary share capital.

> A large and up-to-date factory was erected next to their Preston Works for the production of metallic filament lamps. Work began in 1911 and soon lamps were being manufactured at the rate of 8,000 a day.

The new extension, known as the Britannia Lamp Works, was eventually disposed of between the wars to Siemens Brothers and Co. Ltd. The local trade directory for 1917 shows the Britannia Works to be part of the Dick, Kerr business.

Mixed Pressure Turbo-Alternator Sets:

In 1911 Dick, Kerr & Company supplied a 500 kW mixed pressure steam turbo-alternator to a colliery in Scotland.

The turbine, which was of the impulse type, was designed to give 500 kW with exhaust steam having a pressure of 16 lb. per square inch absolute, or with dry saturated live steam of 90 lb. per square inch, or with a combination of both pressures, and to give the full output when working condensing with a vacuum of 27.5 in., the barometric pressure being 30 in. The turbine was mounted on a combination base and directly connected to its alternator through a flexible coupling. The turbo-alternator was designed to give 625 kVA (500 kW at 80 per cent power factor) at 3000 volts and 50 cycles when running at 3000 rpm. The temperature rise was claimed not to exceed 70 °F above an air temperature of 77 °F after an eight hours run at full speed.

Festival of Empire 1911:

As an aid to the success of the Festival of Empire at the Crystal Palace, a small electric railway was constructed in the grounds by Dick, Kerr and Company from plans by Albion T. Snell, M. Inst. C. E., consulting engineer to the Festival of Empire, who also supervised the work. The line was known as the 'All-Red Route' and was lighted at night by 10,000 incandescent lamps and eighty arc lamps. The route comprised about 1¼ miles of single track laid to 3 ft. 6 in. gauge with 45 lb. rails, with electrical transmission via third and fourth rails. The line was equipped with automatic signals and was divided into sixteen sections. Nine trains were provided. The motor cars were equipped with Dick, Kerr & Company standard type of motor controllers and apparatus.

There was reason behind the name; within the extensive grounds attached to the Palace there were buildings representing all the leading colonies and of India, and representations had been constructed of colonial and Indian scenes and industries, so that the visitor "gets a graphic idea in succession of a number of different countries over which the British flag flies in various parts of the world". To help towards the illusion, as well as to save walking, the plan was adopted of building a light railway, electrically equipped, and constructed in such a course that it would pass in succession through all the exhibits representing the British possessions.

The stations were named respectively South Africa or Pageant Station, Newfoundland, Canada, India, Australia and New Zealand.

The Tramway and Railway World for 6th July 1911 gives a complete and illustrated description of the line.

The London and South-Western Railway

(Reference: The Engineer 24th September and 1st October 1915).

The two articles describe the power house, its architecture and equipment in great detail –

> The power house was situated at Durnsford-road, Wimbledon and adjacent to the river Wandle had a turbine room 257ft long and 55ft wide containing five main and three auxiliary generating sets with space provided for another main unit; the main generating sets which supplied current to the railway were supplied by Dick, Kerr. --- Each turbine was coupled to a 5555kVA three phase 11,000 – 11,500 volt 25 cycle two-pole alternator with direct coupled exciter. The Dick, Kerr turbines were of the impulse type, having one velocity disc with two rows of blades and a suitable number of low pressure discs each with a single row of blades.
>
> All the Dick, Kerr sets were guaranteed to withstand an overload of 25 per cent for two hours and 50 per cent for ten minutes without excessive heating.

The annual report to shareholders reported on the 12th October 1912 appears to have been somewhat downbeat –

> The report highlighted unexpected losses on certain contracts being mainly responsible for the fact that the Directors of Dick, Kerr, and Co., were unable to recommend a dividend on the ordinary shares for the year ended 30th June. The preceding year had shown a profit, after paying debenture interest, of £26,234; for the current year the Board were able to report a profit of only £3,275. Fortunately, a substantial amount of £36,891 was brought into the accounts, making a total sum available of £40,166, to cover 'various fixed charges' and the preference dividend, and leaving £10,016 to be carried forward. The coal and railway strikes had contributed to the misfortunes of the Company.

The Ordinary General Meeting of the company was held at the Cannon-street Hotel on the 15th October 1912, Mr. Claud T. Cayley presiding –

> The Chairman reported that the company "has had a poor year, indeed the poorest year since the foundation of the business. The contracts entered into during the previous five years had been taken in the face of great competition, and as much of the work had been accepted a long time before it could be executed, a considerable number of contracts recently finished had been taken when competition was at its keenest. The recent railway and coal strikes had affected the business at Preston and although stocks of coal were good it had become impossible to get raw

materials in or manufactured goods out of the works. With the works order book at maximum net production during the 12 months under review was the lowest for ten years".

The order book was reported to be 'healthy', the extensions at the Preston works principally in connection with the completion of the lamp works, which was producing for the market.

Two directors had been lost from the board, Mr. John Kerr, the late chairman, and Mr. George Richardson retired from the board for private reasons. The chairman expressed regret that "it had been thought by one or two of their colleagues that this may have been in connection with the poor result of the profit and loss account for the year under review". The chairman assured the shareholders that this was not the case. Mr. John Sampson had accepted a seat on the board.

John Kerr was declared bankrupt on 11th December 1913 and was listed amongst the shareholders until 1914.

Dick, Kerr and Co's Recovery (Times 6th October 1913)

The report of this company of engineers and contractors for the year ended 30th June, which has just been issued, is a much more favourable document than that for 1911-12, although again the ordinary shareholders have to go without a dividend. If it had not been that £36,892 was available from the undivided profits of 1910-11 the company would not have been able, out of the year's results, to pay its debenture interest. The problem was twofold: the strikes had interfered with business and there had been unexpected losses on certain contracts completed during the year.

Certain items in the balance sheet of June 30, 1912, seem to have caused some concern. The item of 'sundry investments' stood at £149,191, almost the same amount as the year before. Beyond the generally known fact that the company held the bulk of the capital in the English Electric Manufacturing Company, it does not appear what these 'sundry investments' are, and nothing is said as to whether the value assigned to them, now that £16,000 has been written off for depreciation, represents a realizable market value or not. Another asset which was vaguely defined was 'capital expenditure at works, including goodwill and patents'. "It would be more satisfactory, and would not injure the company in any way, if these items were separated".

Mr C. T. Cayley presiding at the General Meeting of the company, on 9th October, said that –

"The report showed improved results, although the position was not yet what they would like it to be. They had a large amount of work in hand; in fact they had never been busier than they were at the present time". One shareholder expressed disappointment at "the absence of a dividend on the ordinary shares", and said "it seemed of little use to have the works full of orders if there was no return from it". He asked "how much of the item 'capital expenditure at the works, including goodwill and patents', £611,393 was paid for goodwill". The chairman stated that "£226,000 of the amount was in respect of goodwill and patents". Another shareholder inquired "whether the new American tariff would open up a field for the company under more favourable conditions in the United States". The Chairman replied that "there was still a tariff duty of 30 per cent against them, although it was formerly 45 per cent. That fact made it doubtful whether they would yet do a great deal in America, but it might be hoped that the day would come when they would do so".

Civil Engineering Contracts

As the years progressed Dick, Kerr and Company Ltd. expanded their civil engineering business. They had acquired a considerable experience of the construction business through building tramways, and perhaps anticipating a reduction or at least levelling off in demand for tramway electrification work, it would have made good business sense to utilise their acquired experience elsewhere.

In addition to the civil engineering work associated with the building of steam engines, tram-roads, power generation and railway electrification projects, already mentioned, Dick, Kerr and Co. seem to have been quite prepared to bid for work which was outside their traditional areas of expertise. In the decade leading up to the Great War merging with other companies had reduced competition in the tram-car business thereby helping to maintain profitability in that area. On the other hand, the civil engineering part of the business including tram-road construction was probably starting to slow down, as evidenced by some of the director's reports over this period. It is not surprising that some diversification in the types of contract that the company was prepared to bid for was necessary in order to fill the gaps.

The civil engineering contracts would have been placed with head office, 110 Cannon-street, E.C. and the work undertaken by Dick, Kerr and Company Ltd. at Kilmarnock. It is unlikely that the Strand Road Works in Preston would have had much involvement except, perhaps, as subcontractors, to Dick, Kerr, for electrical equipment. Surviving documents from the period, held in the Lancashire Archive, make no mention of any work done in support of these activities.

Las Palmas:

Before the Great War Dick, Kerr and Company Ltd., had been engaged on a large civil engineering project in Las Palmas, Grand Canary. This involved the laying of a water service to the town of Las Palmas and the use of water falls for the generation of electricity.

Walton Pumping Station:

The pumping station was built to serve the various reservoirs and filter beds in the vicinity, concentrating all the pumping effort into one location. Charles E. Hearson, Chairman of the Works and Stores Committee of the Metropolitan Water Board, laid the foundation stone of the pumping station on 24th October 1908. The building was designed by James Restler (1851 – 1918), Deputy Chief Engineer to the Metropolitan Water Board. The station began operating on 10th July 1911 when John Burns MP, President of the Local Government Board, started the engines.

> Contractor: Dick, Kerr & Co.
> Steam Engines: Thames Ironworks Shipbuilding & Engineering Co.
> Boilers: Babcock & Wilcox, Renfrew.
> Centrifugal Pumps: Gwynnes.
> The steam engines and pumps were replaced in 1926.

National Experimental Tank, NPL Teddington:

A report, in the Engineer for 16th April 1909, on the Spring Meeting of the Institution of Naval Architects includes a presentation from Dr. R. T. Glazebrook, the director of the National Physical Laboratory, being an abstract of a 'Report of the Experimental Tank Committee (1908)'.

The presentation took the form of a brief report on the steps taken by the committee appointed by the Council of the Institution to carry out the scheme for establishing an experimental tank at Bushy (*Bushy Park Teddington*) –

> At a meeting of the Council, held on April 2nd, 1908, a letter was read from Alfred F. Yarrow, Vice President, a marine engineer and shipbuilder, intimating his readiness, subject to certain conditions, to defray the cost, up to £20,000, of an experimental tank to be established in connection with the National Physical Laboratory at Bushy. The proposed tank was to be primarily devoted to research into problems connected with ship propulsion, but was also to be made available, when possible, for conducting experimental investigations of a confidential character for private firms, suitable fees being charged for this work.
>
> Mr Yarrow's munificent offer was gratefully accepted by the Council on behalf of the Institution, and a committee was formed to take the steps necessary to give effect to Mr. Yarrow's proposals.
>
> In order to take advantage of any improvements Dr. Glazebrook made visits to the most recent establishments of the kind, both in the United Kingdom and also in France and Germany; these included Berlin, Dresden (Uebigan) and Paris. The report also provided sectional drawings of each of these facilities in addition to those at Clydebank, Haslar and Washington and goes into some detail on the construction and working of each facility including instrumentation.

On 11th March 1910, the Engineer, reports on the 'rapid progress towards completion of the experimental tank at Teddington' with illustrations –

The report gives details of the form and structure of the tank, the buildings and supporting facilities including workshops. The work was being carried out in accordance with plans suggested by Dr. R. T. Glazebrook, Messrs. Mott and Hay being the consulting engineers, and Dick, Kerr and Co. Limited, Cannon-street, E.C., the contractors for the buildings, tanks, electrical equipment, &c.

> The tank itself was 550 ft. in length with width of 30 ft. between the walls at the top and held 5,000 tons of water, at the south end a shelving beach was provided, on which the waves generated by the passage of the model through the water may break. The tank was constructed of unreinforced concrete throughout except at the docks, the first stage of construction consisted of excavating the ground to a depth of about 9 ft. and forming the concrete bottom. The tank walls are 13 ft. high from the lowest level and the working depth of the water was 12 ft. 6 in. At the top the walls are 2 ft. wide and on them rails weighing 110 lb. per yard were fixed for the carriage. These rails were continuously supported, as it was essential to avoid any sagging, which would give a periodic variation in the velocity of the carriage.
>
> A second and smaller tank was also provided, of reinforced construction, 20 ft. length and 3 ft. 6 in. depth, with an 8 ft. sump at either end. Whereas with the main tank the model was towed along the length of the tank, in the case of the small tank the model was stationary while water was circulated in a continuous stream against it by an 82 brake horsepower centrifugal pump. At a speed of 200 revolutions per minute the pump was capable of delivering 75 cubic feet of water per second. This smaller tank was used principally in connection with experiments on the resistance of plates and shapes of various kinds in a current of water. The tanks included apparatus for producing 1m high waves so that models could be tested in rough or shallow water.
>
> The facility included a wood working and model making shop, from which an overhead crane carried the models to the docks for adjustment &c; a drawing office, machinery room, battery room, boiler room and offices.
>
> The ship models were made of paraffin wax, first moulded to approximately the form required, and then shaped by a special cutting machine to reproduce exactly the lines of the ship. This method of modelling allowed for the shape to be easily modified.
>
> The facility opened on Wednesday 5th July 1911 was originally known as 'The National Experimental Tank', later as the 'William Froude National Tank', after the engineer and naval architect; and finally 'No.1 Tank'. The opening was attended by many distinguished guests; the chair was taken by Sir Archibald Geikie, who as president of the Royal Society was chairman of the general board of the laboratory, and he was supported by Lord Rayleigh, the chairman of the executive committee. Also in

attendance were the Marquis of Bristol, president of the Institution of Naval Architects and Lady Bristol; Mr A. F. Yarrow, Dr. Glazebrook, the director of the laboratory; Sir William and Lady White, Mr. G. S. Baker, the superintendent of the tank; H. E. Senoi Edwards, Sir William Crookes, Mr. R. W. Dana, Rear Admiral Capps, Herr Hillmann, M. Bertin, Sir Norman Lockyer, Mr. Alex Siemens, Sir J. W. Swan, Rear Admiral Moore, Mr. A. B. Kempe, Sir J. Rose Bradford, Sir J. Lannor, Sir Charles Parsons, Prof. Unwin, Sir J. Wolfe Barry, Sir David Gill and Mr. F. W. Balck.

Littleton Reservoir:

Dick, Kerr & Company was contracted in July 1914 to build Littleton Reservoir at Staines, Middlesex. Covering approximately 1⅓ square miles this reservoir for the Metropolitan Water Board was designed to hold 7,000,000,000 gallons of water and would have been the largest reservoir in the British Isles. The contract was terminated in February 1916 by the Ministry of Munitions under the Defence of the Realm Regulations. When Dick, Kerr's disposed of 'certain tools plant and materials' the Metropolitan Water Board took the case to court arguing that 'under the terms of the contract it was their property until completion of the work'. The hearing was concluded on 1st August 1916 with, the defendants, Messrs. Dick, Kerr being –

> 'Restrained from removing plant or receiving money for plant removed except as ordered by the Ministry of Munitions. All the plant which came on to the site became the limited property of the Board under the contract, but in acting under the orders of the Ministry of Munitions the defendants had done no wrong'.

At a further hearing on 19th October 1916 it was held that –

> 'The defendants were not liable either to pay the proceeds of the sale to the plaintiffs (*the Metropolitan Water Board*) or to pay damages for the removal of the plant, so far as they acted as the Ministry's agent, as they were found to have done'.

An appeal on behalf of the Metropolitan Water Board was taken to the House of Lords and on 27th November 1917 the Lord Chancellor, Lord Dunedin, held that –

> 'The prosecution of the works became illegal in consequence of the action of the Ministry of Munitions. It became illegal on 21st February 1916 and remained illegal at the present time. This was not a case of a short and temporary stoppage, but a prohibition in consequence of a war, which had already been in force for the greater part of two years, and would, according to all appearances, last as long as the war itself, as it was the result of the necessity for preventing the diversion to civil purposes of labour and material required for purposes immediately connected with the war'.

The appeal was dismissed, with costs. This became a landmark case setting precedent for later similar cases.

In 1915 Dick, Kerr & Co., Kilmarnock supplied 2 standard gauge 0-6-OT Locomotives in support of this contract.

AMALGAMATION AND THE GREAT WAR

On the eve of the Great War the leading British electrical manufacturers as a group produced less heavy electrical machinery than either of the two leading German firms and less than either General Electric or Westinghouse in the United States. Furthermore not one of the foremost British manufacturing companies had built a major plant for the production of heavy electrical machinery. The leading British producers of power equipment were subsidiaries of foreign manufacturers. British Westinghouse had its major plant in Trafford; British Thompson Houston, a subsidiary of American General Electric, had a major manufacturing facility in Rugby; and Siemens in England manufactured dynamos at Stafford. The leading British owned manufacturers of power-plant equipment were Dick, Kerr & Co. and Parsons; who were mainly concerned with the manufacture of equipment for power plants supplying a traction load.

In Britain the electrical power for railways and tramways was supplied by small power stations dedicated to the needs of the particular project with no apparent attempt to standardise, whereas in the United States and Germany supplies were mainly drawn from the municipal power stations which by necessity required large generating plant.

The financial position of Dick, Kerr & Co. Ltd. in 1914, and reported by the Chairman Mr. C. T. Cayley on 15th October, showed profits earned during the year equalled £44,761; 6% paid on Preference Shares, no ordinary dividend. The need for more working capital was clearly shown –

> In view of the war situation the Directors thought that they would have to tide over a lean year or two in spite of the absence of German competition. The principal contracts in hand were for the London and S.W. Railway electrification, the Lancashire Railway electrification in the Manchester area, including, the first 1,200 V third rail equipment for the Manchester-Bury main line.
>
> Among other contracts were a considerable number of electric tramcar orders, obtained from the United Electric Car Company, who made the bodies. Some of these were for Darwen and for Blackpool, while for the LCC trams a rotary converter was supplied. Among the war-time orders were switchgear, transformers and instruments, for a national projectile factory at Hackney, London.

The Company still made general engineering products and supplied tramway rails to Dover.

During the year 56 motors were made by Siemens Bros. Dynamo Works for Bradford trolleybuses, with Dick, Kerr making the circuit breakers, and the Phoenix Dynamo Company supplying the battery boosters. This was in all probability one of the earliest examples of collaboration between three firms who would, with others, come together a few years later to found the English Electric Company.

A repeat order for a 5,000 kW turbo-alternator constructed jointly by Willans and Robinson and Dick, Kerr was received from Dundee, and for the Neepsend Power Station, Sheffield, an 8,500 kW set of Willans, Dick, Kerr design. Mr. Cayley expressed the opinion that –

> "While undoubtedly some departments would be detrimentally affected by the war others might be expected to derive certain advantages. Concerning the prospects of British manufacturers of electrical machinery, much is being said regarding the benefits to be reaped by the removal of German competition, but it must be borne in mind that the elimination of Germany does not necessarily signify the entire absence of competition. It appeared to him that to whatever extent Germany may be crippled financially for the time being, she would make stupendous efforts to regain her lost markets".

The Profitability of Dick, Kerr & Co. and the Electric Railway & Tramway Carriage Works/United Electric Car Co., 1901-1914

Dick, Kerr & Co. Year to June	Gross Profits (£)	Ordinary Dividend %	ERTCW/UEC Gross Profits (£)	Ordinary Dividend %
1901	69,408*	13	25,121	8
1902	103,758	10	18,895	8
1903	110,449	15	24,190	8
1904	84,170	10	12,345	5
1905	85,007	10	21,824	8
1906	86,354	10	22,497	5
1907	80,476	10	29,066	5
1908	59,558	10	15,269	0
1909	28,168	6	17,042	2.5
1910	22,820	5	16,540	2.5
1911	37,994	5	12,673	0
1912	3,275	0	23,640	5
1913	30,092	0	18,543	5
1914	44,761	0	22,922	5

Note. * English Electric Manufacturing Company only.

It is apparent that Dick, Kerr's strategy of developing a broad diversification of business, from civil engineering to electrical engineering and tramcar manufacture, had made them financially more resilient than their competitors.

At the meeting of shareholders on 21st October 1915, Claud T. Cayley stated that –

"He had no reason to alter the opinion he expressed a year ago that the company would experience some lean years after the war before they could hope for the re-establishment of their standard business, especially in heavy manufacturing". The accounts showed a profit for the twelve months ending June 30 amounted to £46,959-17s-5d., which added to £25,525-7s-11d brought forward from 1914, made £72,485-5s-4d. Out of this debenture interest, trustees fees, and a reserve sum to provide for the premium payable on the redemption of the present debenture absorbed £10,838, leaving a balance of £61,646. Of this total the preference dividend amounted to £18,300, while £25,000 was placed to the reserve, and £18,345 carried forward. The directors were unable to recommend any distribution on the ordinary shares, nothing having been paid on ordinary shares since 1911.

This was followed on 26th October by the shareholder meeting, in London, Chaired by Claud T. Cayley who reported that –

"The work undertaken by the company had been of greater magnitude than the directors had anticipated last year. They had acquired a controlling interest in the shares of Messrs. Willans and Robinson, though it was not proposed to make any alterations in the present directorate and management of that company". He paid tribute to the way in which the staff had worked, and said that "the time lost by the workers at the Preston works from all causes over a period of twelve months was less than 1.4%". "Given reasonable protection of a permanent nature against unfair competition from outside the Empire, there was every reason to look forward with confidence to the future".

A shareholder said "he could not see any evidence that the German control of trade all over the world was being properly dealt with. It was useless to look to the politicians, and he thought the key to the situation lay with the big industrial trust companies of his country, who had provided the money for expenditure abroad".

At the annual meeting in November 1916, Mr. C. T. Cayley reported that –

"For the last financial year to 30th June 1916 the gross profits had increased from £46,900 to £71,000 and the net profits, after the payment of debenture charges, from £36,300 to £60,900. A sum of £25,000 was again transferred to a reserve against contingencies and 6 per cent was distributed on the ordinary shares, on which no return had been made since 1910-11, when a dividend of 5 per cent was forthcoming. Finally £20,000 was carried forward, or £2,000 more than the amount brought forward from the previous year".

"During the year, in addition to extensions of contracts for the supply of war material, which had kept the works fully occupied, the construction, equipment and management of one of the new national factories had been entrusted to the company".

The A.E.G. Electric Company, a subsidiary of A.E.G. of Berlin, was sold by the Controller of Enemy Property to Dick, Kerr and Co.

The Ordinary General Meeting of Dick, Kerr & Company Ltd. was held on 31st October 1917 at the Cannon-street Hotel, Mr Claud T. Cayley, the chairman of the company, presiding –

Comparing the balance sheet with that of the previous year, the net profit showed an appreciable increase. The figures taken in conjunction with the total assets and liabilities indicated the magnitude of the company's operations, although under the prevailing circumstances it was not possible to publish details. The total of the assets shown in the balance sheet was £2,193,688, as compared with £1,753,410 in the previous year.

The increase in the figure of investments was the result of the acquisition of the shares in the United Electric Car Company (Limited). An increase in liabilities under the heading of creditors and estimated liabilities was stated to include the reserve set aside for the excess profits duty.

The Government continued to be the principal customer, with continuously increasing demands both in quantity and variety. All Departments for which the company was working showed their appreciation and confidence by entrusting the company with important work to the limit of its capabilities. The factory erected under the auspices of the company (*probably referring to Hackney*) had been handed over and the erection of two more factories was in hand for another purpose. The acquisition of a controlling interest in the United Electric Car Company (Limited) was seen by the board to be fully justified and they were satisfied that after the war there would be no difficulty in keeping the works fully occupied.

The acquisition of control in Willans and Robinson (Limited) had been consolidated by acquiring 86 per cent of the Preferred "B" stock. Steps were in place to consolidate the organisations of the two companies.

Considerable interest was being shown in the contracting world and among the legal profession in the somewhat celebrated case of the Metropolitan Water Board v Dick, Kerr, and Co. which was still *sub judice*.

As previously indicated it was anticipated that there would be a period of uncertainty during the transition from war conditions to normal conditions of peace.

Among contracts for the year were 7,775 kVA water-driven alternators for export. There were also two 5,000 kW turbo-alternators for the Clifton Junction Power Station for the Manchester-Bury line, where Dick, Kerr had supplied the traction equipment.

Mr. R. H. Prestwich was deputy Chairman. Mr. Rutherford the Managing Director.

In a report headed Engineering Fusions –

Messrs. Dick, Kerr, and Co., engineers and contractors, continue to extend their business by acquiring a financial interest in several manufacturing companies of a kindred character, with a view to ensuring the prosperous development of the business after the war. Some years ago they absorbed the English Electric Manufacturing Company, and they have recently acquired 93 per cent of the capital of Willans and Robinson and also purchased the greater part of the issued Preference and Ordinary shares of the United Electric Car Company. The Preference shares amounting to £100,000 in £1 shares are being offered to the shareholders at par, but the company will retain the Ordinary shares. It is intended that the controlling concern shall act as a central organisation to provide a single selling agency for all the concerns comprised in the group, both at home and abroad. The fusion of interests will enable the companies to avoid the overlapping of foreign branches, and enable new branches to be opened in other foreign centres without any additional expense.

Willans and Robinson

Willans & Robinson was founded, in 1880, as a partnership by Peter Willans and Mark Robinson. The Company took limited status in 1888 and reported in the Engineer for 27th January of that year –

This is the conversion to a company of the business of Messrs. Willans and Robinson, of Thames Ditton, mechanical and electrical engineers, boilermakers, iron & steel founders, launch and shipbuilders. It was registered on the 12th inst., with a capital of £200,000, in £50 shares. The subscribers are:--

* P. W. Willans, Frimley Lock, Surrey, engineer.
* M. H. Robinson, Surbiton, engineer.
* Sir J. C. East, Bart., Maidenhead
* E. L. Scott, Salters Hall, St. Swithin's-lane.

G. J. Douglas, Thames Ditton.

F. C. Peache, Thames Ditton, engineer

Percy A. Low, Long Ditton, engineer

The board of managers is not to be less than three, nor more than seven; qualifications, fifty shares; the first are the subscribers denoted by an asterisk; remuneration £1200 per annum, with an additional £200 for each £10,000 paid-up capital beyond £100,000.

Peter William Willans was a prolific inventor, having taken out some 82 patents. His inventions included the three-crank engine, the central valve engine, forced draft for boilers, electric governors, combination bedplates for generating sets, air condensers, marine reverse gears and magnetic couplings. He also invented the epicyclic reverse gear for marine purposes and for the reverse gear in which the thrust of the propeller is used to keep the members of a friction clutch in engagement.

Willans and Robinson was essentially a boat-building and marine engine building firm from 1880 to 1884, although the three-crank engine was applied to other purposes. In 1884 one of the smaller size launch engines was direct coupled to a dynamo and after full-scale trials over three weeks, at Thames Ditton, a complete generating set including a small boiler was fitted in a parcels van on the London underground railway and electric lighting was tried out. Later the same year four generating sets embodying the same type of engine, with some modifications, were installed in Buckingham Palace where they ran for many years. A similar engine, supplied to Windsor Castle for the royal laundry, was presented to the Science Museum by King George V.

In 1888 Willans and Robinson supplied central valve engines, boilers and other fittings for the sewage disposal works at Kingston-on-Thames. This included two engines of each of 40 indicated horse-power, for driving the mixing mills, drying cylinders, air compressors and other machinery at the works; three pumping engines, each capable of raising 100,000 gallons of sewage per hour, and three – for Surbiton – each able to raise 45,000 gallons per hour – eight engines in all.

Perhaps one of his most significant innovations was the concept of the interchangeability of manufactured components. By 1891 the manufacturing limits of the central valve engine from the smallest to the largest size built were bound into a book. These limits with some modifications held good until the building of the central valve engine ceased after the 1914-1918 war.

Peter William Willans (born on 8th November 1851) died on 23rd May 1892 whilst driving his dog-cart from home to Brookwood Station. The horse took fright at a train, bolted, and he was thrown out of the trap receiving a fatal head injury.

In March 1894 the private limited company was wound up and Willans and Robinson Limited were incorporated with Mr M. H. Robinson, Sir Gilbert Clayton East and Captain M. H. P. R. Sankey, as directors.

By 1895 it was becoming clear that the Thames Ditton works was unlikely to meet growing demand for the company's products and the board made the decision to relocate production of its large engines to a new engineering works on Newbold Road, Rugby; plans for the first phase being presented to the Urban District Council in April of that year. The Victoria works included provision for a machine shop, forge, foundry, pattern shop, testing department and general offices in separate buildings. The works also had sidings from the Leamington branch of the London and North-Western Railway. To Thames Ditton was assigned production of the smaller sizes of central-valve engines, some components for larger sizes but chiefly, the launch engine and boat building work.

By 1897 the new works, with the exception of the foundry, was in production and in March of that year the first Rugby-made central-valve engine was in hand.

At this time many Willans central-valve engines were being coupled to dynamos made by Dick, Kerr and Co. and Siemens Brothers, both of which firms would become part of the English Electric Group.

Three years after the opening of the Rugby works, a Willans and Robinson engine was shown at the Paris Exhibition. Regarded as an outstanding example of engineering, it was awarded a gold medal.

In 1902 an agreement was reached with C. A. Parsons and Co. for the license production of steam turbines to their design. The work was placed in the hands of Mr. E. G. Izod, then Assistant Test Engineer, who successfully brought into being the steam turbine business at the Rugby Works.

(*These days Mr. Izod is probably best known as the inventor of the Izod impact-testing machine*).

Between 1903 and 1904, 13 steam turbines had been ordered, and in 1906 four 1000 kW 1,500 rpm turbines constructed for South Tees were subjected to a rigorous seven-day test.

In 1904 negotiations with the Diesel Engine Co. of London resulted in a license agreement for the manufacture of diesel engines from drawings made by Messrs. Sulzer.

During the Great War output was progressively switched to armament production and in addition to adapting its engines and traditional skills to the war effort the company cast bronze propellers in large quantities for the Navy.

In 1916 Willans and Robinson, lacking a vigilant and prudent financial policy, was consolidated into the Dick, Kerr organisation.

Munitions

The outbreak of the Great War found Britain totally unprepared for land hostilities on a Continental scale. The Army mainly used for policing the widely scattered Empire, was a small, highly trained force of professional soldiers who lacked both the numbers and the equipment for large-scale operations against European armies.

Although Britain was a leading manufacturing country, its industries were almost entirely concerned with peacetime production. We were deficient in machinery for the production of rifles, machine guns and heavy artillery. Our heavy artillery for example, consisted mainly of 6-inch howitzers of obsolescent design, some 6-inch guns mounted on coastal forts and one experimental 9.2-inch howitzer. At this time the War Office was supplying only shrapnel for the field guns and of the shells even for field howitzers and 60-pounders, 70% was shrapnel. The single 9.2-inch howitzer sent out to France in October 1914, fired high explosive, but only had a limited supply of ammunition.

At the end of 1914, the British Army Expeditionary Force had an acute shortage of heavy guns and an inadequate supply of both light and heavy ammunition. In August 1914 the state-owned ordnance factories were providing the Army with about a third of its weapons and at this time there were only sixteen firms tendering for War Office munitions contracts.

To address this problem the first 'Defence of the Realm Act' (DORA) was passed on 8th August 1914, giving general powers to the government to make regulations for the conduct of affairs under war conditions. A second bill, laid before Parliament on 28th August 1914, to extend these powers and including control over armament factories and their workers. On 9th March 1915, a Bill was laid before the House of Commons to amend the provisions of the Act to extend the power of control to enable the Government to take over and use any factory or workshop.

The Prime Minister, who released details of the new Coalition Government to the press on 26th May 1915, also announced the following: "The Prime Minister has decided that a new department shall be created, to be called the Ministry of Munitions, charged with organising the supply of Munitions of war. Mr. Lloyd George has undertaken the formation and temporary direction of this department and during his tenure of office as Minister of Munitions will vacate the office of Chancellor of the Exchequer." The quarters allocated to the new Ministry were at 6 Whitehall Gardens, London, the former residence of the artist Lockett Agnew. Lloyd George appointed successful businessmen to the chief executive posts and by the end of his year of office, he had transformed the British economy. At the end of the war, the Ministry was employing a staff of 65,000 and had over three million workers under its control.

The materials which the Ministry of Munitions brought under its control involved nearly 100 main categories, and included not only the obvious munitions such as ammunition and weapons and raw materials such as iron and steel, but also included plaster slabs, gas masks, waste paper and boxes. Ultimately, the Ministry assumed responsibility for all supplies of materials; it controlled distribution of raw material to non-munition as well as munition trades, thereby bringing all industries together involved in the production of munitions (either directly or indirectly) under the control of one department. This legislation made it possible for the Ministry of Munitions to recruit private firms such as Dick, Kerr and Co. and to place contracts with them for the manufacture and supply of Munitions.

Walter Rutherford was amongst the first to realise the part that the engineering industry would play in a successful outcome. In the first few weeks of the war his plans were made and the company agents were buying machinery at home and in the United States. The first shells were being produced at the Preston works before the end of 1914, and within 15 months the whole factory was on munitions and the weekly output had risen to 500 tons. The first batch of shell was bonded on 3rd March 1915.

James Conner with the assistance of Robert Livingstone had the task of re-equipping the Strand Road Works with the necessary machine tools for the manufacture of projectiles.

The extent to which Dick, Kerr and Co. became involved is evident from the agenda book for the period 1915-17 and the Secretary's Private Letter book for 1916 which give us a glimpse into the activities of the company during the war years. Information is, otherwise, hard to come by, presumably due to security restrictions at the time.

27th January 1915, an order is received from the War Office for 2,500 cartridge boxes.

30th June 1915, orders are received for 50 pontoons for the War Office and 1 electric locomotive for Harland and Wolff.

27th October 1915, the War Office ordered 100 general service wagons.

2nd January 1916, correspondence with the Blackburn Munitions Board with regard to a contract for shells.

16th January 1916, a continuation of the existing shell contract is agreed at schedule prices beyond the 30,000.

15th May 1916, 2,016 shells are put into bond, representing one week's production, bringing the total to 15,211 shells to date. Also, an enquiry was received for 200 pontoons and an appointment was made to see Col. Robertson at Woolwich.

There is a need to increase the capacity of the central bond and Dick, Kerr are requested to clear, as quickly as possible, the space at present occupied by them.

16th May 1916, an enquiry is received, dated 11th May, from the Director General of Munitions Supply to increase the number of cars on order to 100.

28th June 1916, a production problem is highlighted with several thousand shells having eccentric walls, 1,000 shells had been rejected. Blackburn Munitions Board agreed to extend the time for putting 30,000 shells into bond.

26th July 1916, Lincoln Chandler, representing the Saltley Works of the Metropolitan Amalgamated Railway Carriage and Wagon Co. agrees to accept a further contract from Blackburn Munitions Board for 600 shells per week.

In August the Company invested £10,000 in war stock at ½%.

3rd August 1916, Shell contract statements are prepared for Sheffield, Blackburn Munitions, including shell noses, and the Foreign Office.

13th September a report to R. B. Barningham informing him that of the 184 employees who had joined the army 5 have been killed or are missing, 3 are prisoners of war and 9 have been wounded. *Names were not given*.

One of those who gave his life is commemorated in the Manchester Grammar School Magazine for October 1917, listing ex-Students who had been killed in action including: Lieut. Henry Butler, after leaving the school in 1903, was on the staff of Messrs. Dick, Kerr & Co., of Preston. He enlisted in the King's Liverpool Regiment in August, 1914. In September, 1915, he obtained a commission in the 11th Battalion Yorkshire Regiment. In June, 1916, he transferred to the R.F.C. He was engaged as pilot in long reconnaissance work till his death in action on March 25th , 1917.

2nd October a report to R. B. Barningham, at Trinity Mansions Hotel, Eastbourne, that 1,000 shells had been put into bond the previous week and that Mr. Rutherford was quoting for the supply of huts, 80ft x 15ft for erection at three aerodromes.

Also, correspondence regarding a new shell contract, the old shell contract, shell noses (Blackburn Munitions Board), filling of shrapnel shells and the Moscow enquiry.

11th October, it was reported that - yesterday ran out of steel for shells, three days' supply arrived this morning, do not know when we shall get any more. The whole district is only getting about ⅓ the amount of steel needed.

1st November, R. B. Barningham was informed that with regard to the 18 pounder rejected shells the Ministry were prepared to purchase at £5-10-0 per ton and that it was proposed to reject the offer since there 'was a chance' of obtaining £10 per ton from Hadfields.

6th November, correspondence with the Ministry of Munitions, Manchester regarding letters dated the 10th and 17th of October and the shell contract. An order was received from Dick, Kerr and Co. for car bodies.

4th December 1916, and further correspondence with the Blackburn Munitions Board regarding the shell contract.

3rd March 1917, interim cheques are approved – Ministry of Munitions £6-16-0 and Dick, Kerr and Co. £50,000-0-0.

From the above correspondence the initial production rate for shell appears to have been in excess of 2,000 per week increasing to 2,600 from July 1916. The supply of steel for the manufacture of shell seems to have been a problem from as early as October 1916 and quality of manufacture was a problem with large numbers of shells being rejected at an early date.

By the end of the war the output of projectiles had reached 30,000 a week, and totalled during the war period, 3,300,000 of all calibres, including both high explosive and shrapnel.

The range of projectiles and other artillery related products manufactured during the war included –

- 60 Pounder Shrapnel Shell.
- 13 Pounder Shell.
- 13,5 inch Naval Shell.
- 10 inch Bombs.
- Large calibre Shell for the Italian Navy.
- 6 inch (2 cwt) Towed Medium Howitzer Jackets.
- 7,5 inch Howitzer Barrels.
- Breech Rings, Trunnion Rings and other Gun Parts.

In producing this immense output 94,000 tons of steel, 1,900 tons of copper and 15,000 tons of lead were used.

The large lathe equipment was engaged on turning and wire winding 6 inch gun tubes, machining of 6 inch Howitzer Jackets, 7,5 inch Howitzer Barrels, Breech Rings, Trunnion Rings, and other gun parts. A special plant was installed for re-tubing Guns, and 165 6 inch Guns were renewed in this way. 230 6 inch Howitzer Gun Bodies were built and completed, and in addition, 96 7,5 inch Barrels and 110 6 inch Jackets were machined to help out other contractors.

Strand Road, East Works – Shell filling C1915. 60 Pounder Shrapnel Shell.

Strand Road, East Works – Shells in Production C1915. 13,5 inch Naval Shell.

13 Pounder Shell Casings await Filling.

Gun Barrel refurbishment. 6 inch (2 cwt) Towed Medium Howitzer. The Great War – Munitions Manufacture at the Strand Road Works of Dick, Kerr & Co. Ltd. C1915.

Shell Production. Main Machine Shop, West Works.

National Projectile Factory

From 1916 Messrs. Dick, Kerr & Co. were contracted by the Ministry of Munitions for the equipping of a National Projectile Factory at Hackney. In 1917 the board reported that one National Factory has been handed over to the Government, another two are being built by us for the Government (there is no evidence that these two factories were completed).

The War Department's National Projectile Factory at Hackney was opened in 1916 on part of the Marsh near the G.N.R's sidings and the Hackney cut. The factory was on the site known as Mabley Green; it was pulled down in 1919.

Army and War records provide the following information –

NPF Hackney Marshes, London.

Contractor: Dick, Kerr & Company Ltd.

Opened: February 1916.

Area: Land owned by London Corporation Council.

Management: Dick, Kerr & Company Ltd.

Munitions: 6 in. HE steel (chemical) and 6 in. cast-iron (HE).

Robert Livingstone was transferred to Hackney with responsibility for general machine layout and installation of the 'requisite tooling'. He returned to the Preston Works in 1918 to resume his duties with the recently formed English Electric Company, as Shop Superintendent. In 1920 he became Works Manager when James Conner was transferred to London.

War Department Light Railway

By the middle of August 1916 it was clear that the transport arrangements for the Somme offensive had been inadequate and the War office took such a serious view of its consequences that Sir Eric Geddes, then deputy general manager of the North Eastern Railway, was asked to visit France and report on the transport situation in the British area. His report made it quite clear first, that some form of overall organisation was needed to plan and co-ordinate military transportation and, secondly, that a vast expansion of all types of transport was necessary.

In consequence, and from 5th October 1916, the Ministry of Munitions took complete responsibility for the design, manufacture and supply of all railway material for military purposes and both the Ministry and War Office agreed to this proposal.

The first task was to put into effect plans for the provision of railway equipment, both standard and narrow gauge. The outline scheme, which followed on 9th October included:

Track (20 lb/yd including curves and sidings)	1,000 miles
Electric locomotives	200
Overhead electric equipment	200 miles
Movable power-stations for same	20
Portable workshops for electric engines	10

The original plan was to use steam locomotives exclusively for transportation to the front lines, and electric equipment in the more exposed areas near the front lines where the noise and smoke of steam locomotives might attract the enemy. The success of the first batches of Simplex petrol tractors to arrive in France resulted in the cancellation of the electric overhead equipment and generating station, and the substitution of tractors for some of the steam locomotives originally envisaged in the programme. Conditions in the trenches during 1917-18 proved this to have been a wise move since the maintenance and camouflage of overhead wiring in the forward areas would have been impossible.

In order to provide more flexibility of choice the War Office decided to have 200 electric locomotives built as self-contained units in the form of petrol-electric tractors. These locomotives were for use on the 60cm railways being built for transporting munitions and other supplies to the front line.

Orders were placed in November 1916 for 200 4-wheel 40-45 hp petrol-electric tractors, 100 from British Westinghouse Ltd. and 100 from Dick, Kerr & Co. Ltd.

Each order was regarded as being for 50 pairs of tractors, the 'A' and 'B' units of each pair being different in that one had provision for trolley-pole equipment. The trolley poles were not fitted but the sockets and motor connections were, and the electric motors were so arranged that they could be driven direct from a suitable current supply. Furthermore, the generators of each pair could be connected in parallel through a built-in junction box. A major requirement was that they should be able to generate full power for long periods when stationary without the radiator temperature exceeding 70 °C.

The initial design was undertaken by the British Westinghouse Co., of Trafford Park, to whom the first order for one hundred units was given, and they in turn, sub-contracted non-electrical parts to outside firms.

The first three completed by Dick, Kerr were tested at Dinas Junction on the North Wales narrow-gauge railway and it is believed that they then went to the War Department testing and training depot at Longmoor, Hampshire. The other ninety-seven were shipped to France, together with the further one hundred which had been ordered from British Westinghouse.

A Pair of Dick, Kerr Petrol-electric Locomotives built in 1916 to transport Munitions to the front line in France Made by the United Electric Car Company Ltd. Strand Road, Preston.

There has been some confusion over the actual number of tractors delivered by Dick, Kerr; the earliest published record from the Company was in 1919 in a programme for the Peace Day fete held at Ashton Park, Preston on 2nd August 1919 which states –

> We undertook a very urgent order for the War Office for the supply of 100 locomotives for use on the Western front to haul ammunition and other supplies from the railhead to the front lines. These Tractors were petrol electric type, steam locomotives being unsuitable on account of the conspicuous targets they made for enemy gunners.

'Electricity in Transport' published by the English Electric Company Limited, in 1951, claims –

> Dick, Kerr & Company Limited supplied ninety small petrol-electric locomotives for service in France on a 2-ft. gauge track equipped with 40-h.p. petrol engines, Phoenix (Bradford) generators and two Dick, Kerr motors.

W. J. K. Davies in 'Light Railways of the First World War' quotes War Department records confirming that 'one hundred were delivered'. 'They were very similar to the British Westinghouse ones but were built entirely by Dick, Kerr's except for the generators built by Phoenix, and the petrol engines, supplied by Dorman's, and were slightly less powerful'. 'They could be distinguished externally in having louvered instead of panelled bonnets and by differences in the frame design'.

The final drive had a very low gear ratio so their greatest fault was their slow speed, nine-notch tramway type controllers were used, equipped with a position for electric starting of another tractor, a very useful feature since the big engines were very hard to start by hand, and quick results could be achieved by using the other tractor's generator as a starter motor for the petrol engine. The vital parts were armoured against small arms fire, the armour being detachable.

There was also the facility to use the locomotives as mobile generators supplying 500 volts, and they could, therefore, be used to power up one or two of the unpowered vans of the mobile workshop train or a mobile sawmill.

Flying Boats

In the latter part of 1917 orders were placed with a number of manufacturers for Felixstowe flying-boats. By October 1918 the R.A.F would have a total of 53 F.2As, and 96 F.3s on charge. All the F.2As and 29 F.3s were at seaplane stations in the United Kingdom; of the remaining F.3s, 18 were attached to the Grand Fleet, 13 in the Mediterranean, one was at an Aeroplane Repair Depot, 18 were with contractors, and 17 were in store. Ordered from Dick, Kerr and Co. as F.3s were those with serial numbers N.4100 to N.4159 and N.4230 to N.4279.

Wings and tail units were manufactured by the United Electric Car Co. and married up to hulls from other manufacturers, including, Boulton and Paul of Norwich and Borwick & Sons a boat and floatplane manufacturer on Windermere.

Felixstowe F3. Wing Structure Assembly. United Electric Car Company Ltd., Strand Road, Preston.

Felixstowe F3 Production, Wing Doping. Dick, Kerr & Co. Ltd. at the New Aircraft Works in Lytham 1918.

Dick, Kerr & Co. Ltd. Lytham Works 1918. Felixstowe F3 Flying Boat Fuselage Assemblies.

Felixstowe F3 from the Second batch of 50. Complete and ready for despatch.
Dick, Kerr & Co. Ltd. Lytham Works 1918

The completed units were then stripped down for transport from the Preston Factory to South Shields on Tyneside for reassembly and flight testing. 35 flying boats had been manufactured and tested in this way when in 1918 it was decided that it would be more efficient to have a facility closer at hand and a factory was built on land at Lytham with a slipway into the Ribble estuary. Dick, Kerr received a further contract for F.3s and some F.5s of which only 11 F.3s and 2 F.5s were completed before the armistice.

Traction Business

Throughout the war Dick, Kerr and Company continued with the manufacture of traction related products, since after the war they would need to re-establish their pre-war business as quickly as possible. Electrical equipment and tramcar manufacture was continued for the duration, although at reduced output bearing in mind that war work took up most of the works manufacturing capacity.

Tramcars

On 16th May 1916, R. B. Barningham arranges a meeting with Mr. Johnson of the Brush Co. to discuss the 'Moscow' enquiry, tramcars for the Municipality of Moscow, Brush along with Westinghouse were the agents. *It would appear that, despite the pressures of the war effort, there is still time to discuss the export of cars to Moscow.*

Correspondence regarding this enquiry continues through the year, a British Ministry of Munitions Class A Priority Certificate would be required. Funding for this contract was provided for by short term renewable bonds in pounds Stirling redeemable one year after declaration of peace with interest at 7%. The value of the order was for £140,000.

26th May 1916, orders are received from Melbourne, 5 trucks; Salford, 1 wagon and South Lancashire Tramways, 6 top covers.

22nd June 1916, orders are received from Nelson, 5 cars at £450 each and Sunderland, 1 car at £590.

26th July 1916, an enquiry had been received from Les Ateliers Metallurgiques in Belgium for the supply of 3,200 seats for cars being built in occupied territory; this is referred to HM Secretary of State for Foreign Affairs in Downing Street.

18th October, new orders received included, Nottingham, 6 car bodies at £423 each; Rotherham, 2 car bodies at £439 each and Blackpool 6 pairs of bogie trucks at £186 per pair.

During the period 1915 – 18 some 27 complete cars were produced, for 6 customers, top covers and miscellaneous equipments were also supplied.

Generating Plant

At the end of 1917 an important contract for a generating plant was awarded to Dick, Kerr & Company Ltd., by the Glasgow Corporation for a new power station in the course of erection at Dalmarnock. The contract included a turbo-alternator set, with condensing plant and step-up transformers designed for a continuous output of 18,750 kW. The turbine was a single unit standard Dick, Kerr impulse type machine, arranged to run at 1500 rpm on steam at a pressure of 250 lb. per square inch at the stop valve, superheated to 650 °F. The turbine was to be direct coupled to a 6,500 volt, 25-cycle three phase Dick, Kerr alternator, having a maximum continuous rating of 24,000 kva at 80 per cent power factor.

Relief Fund

To aid the families of men who 'joined the Colours', a relief fund was inaugurated in August 1914. Subscriptions were given weekly by the employees, and to the total a similar amount was added by the firm.

The administration of the fund was carried out by a Committee elected from the Office and Shops. This Committee ably carried out their 'somewhat onerous duties', the work involved was considerable, as the sum of over £24,000 was distributed in Relief altogether. The weekly donations rose from 50 in September 1914 to 370 in September 1918, and the Committee went very carefully into the merits of each case before any grant was authorised. The total number of Families who had benefited by the fund was 604.

Brass Band

A Works Brass Band was formed in January 1918, and placed itself at the disposal of the War Organisations of Preston, with the result that the Military Hospitals, Station Buffet and the Hospital for Disabled Soldiers benefited to the extent of £740.

The Dick, Kerr Ladies' Football Team

Ask almost any person in the North West and particularly in the Preston area, what do you know about the Dick, Kerr Company? The answer is as likely as not to be, the Ladies Football Team.

The Dick, Kerr Ladies football team owed its existence to a man by the name of Alfred Frankland. Frankland was born in 1882 in Freckleton, Lancashire to Thomas Frankland and his wife Jane (nee Sumner) and attended school in Freckleton. By the time Alfred was aged 18 the family had moved to Preston and Alfred was working as a Clothiers Shop Assistant. In August 1905 Alfred marries Margaret Bradley and at the time of the 1911 census they are living in Chorley and Alfred is a Clothing Shop Manager.

With the onset of war, the sudden increase in manufacturing workload and pressures brought about by the receipt of large contracts from the Ministry of Munitions, combined with the loss of experienced male staff to the armed services, resulted in the urgent need to recruit a large number of new and largely inexperienced personnel. Large numbers of women were recruited for the manufacturing workshops but an increase in clerical staff was also essential if the contracts were to be progressed efficiently and effectively to meet the tight timescales imposed by the War Department. Alfred Frankland was amongst those recruited for clerical duties.

Alfred had noticed the munitions girls kicking a ball around during their breaks. The women had already played one match against the youths of Dick, Kerr's when Alfred started taking an interest, and he first became involved when a plan was hatched to play a bigger match on Christmas Day 1917, in aid of the local hospital for wounded soldiers at Moor Park. Alfred Frankland offered to help and he booked Deepdale, the home of Preston North End Football Club.

The local press covered the game and reported in some detail under the headline Ladies' Football, 10,000 Present at Preston Match –

> There were 10,000 people at Deepdale yesterday afternoon at the ladies football match between teams representing the munition workers of Dick, Kerr's and Coulthard's, and the appearance of the historic enclosure was quite like old times. The proceeds from the sale of tickets, which were supplemented by a collection taken by wounded soldiers on the ground, were for Moor Park Hospital, and when all the money has been got into the funds the institution should receive a handsome windfall. – The players wore orthodox football costume, jerseys, dark blue shorts

> and regulation boots. Coulthard's were in red and white stripes and Dick, Kerr's in black and white, with the addition of natty, close fitting hats to match. Corsets were barred. So far as appearances went, Dick, Kerr's seemed to hold some advantage between two athletic-looking sides. – After the Christmas dinner the crowd were in the right humour for enjoying this distinctly war-time novelty. There was a tendency amongst the players at the start to giggle, but they soon settled down to the game in earnest. Dick, Kerr's were not long in showing that they had suffered less than their opponents from stage fright and had a better all-round idea of the game. Woman for woman they were also speedier, and had a larger share of that quality which in football slang is known as 'heftiness' --- Their forward work was often surprisingly good, one or two of the ladies displaying quite admirable ball control. At first the spectators were inclined to treat the game with a little too much levity, and they found amusement in almost everything from the pace, which until they got used to it, has the same effect as a slow moving kinema picture, to the 'how-dare-you' expression of a player when she was pushed by an opponent. But when they saw that the ladies meant business, and were 'playing the game' they readily took up the correct attitude and impartially cheered and encouraged each side. Within five minutes Dick, Kerr's had scored through Miss Whittle, and before half time they added further goals by Miss Birkins – a fine shot from 15 yards' cut just under the bar – and Miss Rance. Result – Dick, Kerr's 4, Coulthard's 0. The game raised £488-7s-0d for the Moor Park Hospital.

On the day 10,000 people poured through the turnstiles to the accompaniment of the works' brass band.

Alfred Frankland could not have achieved this or the later successes, during and after the war, without the official sanction of the company, and the Management gave their full support. The club went on to gain national and international acclaim, playing international games, visiting France in 1920 and the United States in 1922 and playing against visiting French teams on a number of occasions in the early 1920s.

On 22nd November 1918, and reported by the Lancashire Evening Post –

> Though the armistice has been signed, and there is no longer the same need for munitions, the committee and players have decided to carry on as far as possible with their fixtures, as the local charities are in as great a need of help as before, and an interesting game should be seen at Deepdale to-morrow, when a really good side from the British Westinghouse Co's works at Manchester will provide the opposition. Dick, Kerr's team will be: - Clayton, Redford, Crosier, Birkins, Kell, Booth, Walker, Standing, Mitchell, Dickinson, and Jones.

The Dick Kerr Ladies Football Team 1917.

The Dick Kerr Ladies Football Team 1919.

Dick, Kerr Ladies v Rest of England. By Floodlight on 17th December 1920.
Watched by 10,000 spectators, Dick, Kerr's won 4-0.

Moor Park Hospital for Wounded Soldiers 1916.

Reported on 14th September 1920 –

> Arrangements have been made whereby the Dick, Kerr's Ladies Football team will play a match in London against a Lyons' Ladies team, on one of the League grounds, on November 9th, on their return from the tour in France. An offer has also been made to meet the Bath side at Blackpool during the present month, in aid of the Red Cross and the Manchester and Liverpool Universities Appeal Fund.

Thursday 16th December 1920 was the date for a particularly interesting game of football held between Dick, Kerr Ladies and the 'Rest of England' at Deepdale –

> The game was unique in that it took place in the evening with the ground illuminated by searchlights and the ball was painted white. The crowd of 10,000 plus spectators witnessed a picturesque scene; two military searchlights, lent by the War Office, one at either end of the ground; three powerful beams of light above the playing area, around which and close to the touchline were placed 40 carbide flares, also lent by the War Office. A total of £600 was raised for the Ex-Service Men's Distress Fund. The result of the match was Dick, Kerr's 4, Rest of England 0.

By 1921 the team's popularity was at its height, they were the team that everyone wanted to see and they had been booked to play an average of two games a week all over the British Isles. They played over 60 games of football that year while still working full time at the factory, and they were watched by almost 900,000 people throughout the country.

In 1921, following complaints that, for a number of reasons, football was not considered a suitable sport for women, the FA banned the girls from using league grounds and effectively changed the course of women's football forever. But in spite of the prejudice of the Football Association the Dick, Kerr Ladies went on to play over 800 games of football at home and abroad and they made over £180,000 for charity.

Some interesting facts:

- They were the first women's team to play in an organised match wearing shorts.
- They were the first ladies team to play an international match, hosting France in 1920.
- They became the first British women's club to embark on an overseas football tour, a reciprocal tour of France in 1920.
- On 17th December 1920 they were the first women's team to play a match at night using a white football. They approached the War Office for the loan of two anti-aircraft search lights to illuminate the pitch and permission was granted for their use from Winston Churchill.

- They can claim the biggest crowd ever recorded for a women's club team game. On Boxing Day 1920, 53,000 spectators packed into Goodison Park, Everton, for a match against St. Helens Ladies, the Dick, Kerr Ladies won 4-0.
- The team can boast the all time leading goal scorer in women's football. Lily Parr played for the Dick, Kerr Ladies from 1920-1951 and she scored over 1000 goals during her football career; only the legendary Pele has scored more goals than Lily.

Dick, Kerr's Darktown Minstrels

The Lancashire Evening Post for Tuesday 15th May 1917 reported –

For Discharged Soldiers – A successful concert was given on Saturday night in the Lancaster-road Congregational Schoolroom, Preston, by Dick, Kerr's Darktown Minstrels a well-qualified troupe, under the conductorship of Mr. A. K. Turner, and Mr. E. S. Edmundson at the piano. At the close of the programme Mr. G. R. Palmer, secretary of the Preston and District Discharged and Disabled Soldiers' and Sailors' Association, thanked the audience for their patronage, and explained some of the objects of the association on behalf of which the entertainment had been promoted.

In late October 1917 at a Sailors and Soldier's Free Buffet, a report on donations received during October for the 'cost of upkeep' of Moor Park Hospital included – Aircraft Workers, Messrs Dick, Kerr & Co. Proceeds of Concert organised by Minstrels from Messrs Dick, Kerr & Co. £5-0-0.

Other concerts given by the troupe, and reported on, during 1917/18 included –

St. Joseph's Hospital, Mount Street, Preston; Public Halls, Preston and Fulwood Workhouse.

Employees Killed in the Great War

Men from Dick, Kerr's Preston Works killed in the Great War totalled 149. They were commemorated in a Peace Day brochure produced by the English Electric Company in August 1919; the list of those who gave their lives is reproduced in appendix 1.

Preston Trade Directories:

A brief chronological history of the Strand Road Works, prior to 1918 gleaned from the local Preston Trade Directories, and including the period before Dick, Kerr ownership, is of interest if only to confirm some previously mentioned facts but also to add a few names to the story.

1869	**West Strand.** The North of England Carriage & Iron Company (limited), Edwin Quadling, Manager and Secretary.
1873/1874	**West Strand Road.** North of England Railway Carriage & Iron Co., R. C. Heslop, Esq., M.D., Chairman Job Richards, Managing Director Joseph H. Bennett, General Manager Thomas Howarth, Secretary.
1880	**West Strand.** North of England Carriage & Iron Co., Ltd.
1905	**Strand Road East.** Electric Railway & Tramway Carriage Works Limited. L. E. Pickin, Secretary. **Strand Road West.** Dick, Kerr & Co., Limited. Electric Tramways & Railways Contractors.
1917	**Strand Road.** Dick, Kerr & Co., Electric Railway Engineers. Dick, Kerr & Co., Lamp Manufacturers, Britannia Works (Watery Lane).

ENGLISH ELECTRIC

The Ordinary General Meeting of Dick, Kerr, and Co. (Ltd) was held on 14th November 1918 at the Cannon-street Hotel, Mr. Claud T. Cayley, the chairman of the company, presiding.

It was reported, the policy of the directors had been gradually to prepare where possible, for the reconstruction of the business on post-war conditions.

The profits for the twelve months ended June 30th 1918, equalled £87,916. The works had been fully occupied, but the United Electric Car Company's Works had been extended. A dividend of 10% was paid on Ordinary shares. Orders during the year included turbo-alternators and transformers for Glasgow, and a considerable amount of tramway equipment. The total on the asset side of the balance-sheet was over £2,250, 000, a record figure. Investments stood at £338,000, as against £341,000 in 1917. The decrease, however, was the net result of two large transactions, the acquisition of the bulk of the 'B' Preference stock of Willans and Robinson (Limited), partly by cash and partly by exchange of shares which were issued for that purpose. On the other hand, the whole of the holding of United Electric Car Company (Limited) had been disposed of. The work of consolidating into the organisation those of Willans and Robinson and the United Electric Car Company had been completed and the management strengthened by the appointment of Mr. Davenport and Mr. Bennett, from Rugby, and Mr. Conner, a long standing member of the staff. An alliance had been made with Siemens, Brothers, and Co. (Limited) the arrangement being consummated by an exchange of directors, Mr. Mure Ritchie from Siemens and Claud T. Cayley from Dick, Kerr's. There was no competition in manufacture as Siemens Brothers & Company made cables, dynamo machinery, telephones, etc., while Dick, Kerr & Company specialised in apparatus for railways and tramways, steam turbines and the larger classes of electrical machinery.

It had also been decided to establish companies in France and Japan to exploit Dick, Kerr's manufacturing rights in connection with apparatus for railways and tramways. In the belief of the chairman "there would be a large after-war demand for all classes of electrical machinery, especially large-sized steam turbines, and he assured the shareholders that the company's facilities for taking care of this demand would be ample".

It had been brought home to the management that the manufacturing capacity of the country for heavy electrical machinery was much below demand, "and that in our own case we must, if we wish to retain the position we hold, at present, find facilities not only for increased output, but also for production of individual units larger than our present shops admit of". The arrangements which required Treasury sanction involved the Coventry Ordnance Works (Limited).

> An option agreement had been entered into with the Coventry Ordnance Works (Limited) which would involve an exchange of shares with those of a new company, which in the event of an offer being accepted, would control both concerns; full particulars were promised in due course. All the ordinary shares of the Coventry Ordnance Works were held, 50 per cent by Messrs. John Brown and Co. and 25 per cent., each by Cammell Laird and Co. and the Fairfield Shipbuilding and Engineering Company ---
>
> The shops laid down for making heavy armament are equipped with the most modern and efficient machinery, perfectly adapted for making large electrical plant, for which a large demand is expected. The acquisition of the Coventry Works will relieve Dick, Kerr and Co. from the necessity or otherwise of undertaking works extensions which would involve considerable delay.

Dick, Kerr and Company Expansion

A report in the Electric Railway Journal dated 1st February 1919 –

> The firm has established in France and Japan companies to exploit its manufacturing rights in connection with apparatus for railways and tramways. The affiliation with the Coventry Ordnance Works, which is the latest 'deal', is very important. These works, it is expected, will be devoted to the production of heavy electrical machinery. The Coventry Ordnance Works is not a public company, but is owned by the world renowned firms of John Brown & Company, Cammell Laird & Company and the Fairfield Shipbuilding & Engineering Company. In pursuance of the expansion policy there was registered in December a new concern called the English Electric Company, Ltd., with a nominal capital of £5,000,000 in £1 shares (£1,500,000 preference and £3,500,000 ordinary). The shareholders of Dick, Kerr & Company and of the Coventry Ordnance Works will exchange their holdings for shares in the new company. The names of the first directors of the new company are instructive and significant. They are C. T. Cayley, chairman of Dick, Kerr & Company; B. A. Firth, chairman of Thomas Firth & Sons; W. Rutherford, managing director of Dick, Kerr & Company; J. Sampson, director of John Brown & Company; P. J. Pybus, managing director of the Phoenix Dynamo Company; Col. J. H. Mansell, managing director of the Coventry Ordnance Works; A. Gracie, Fairfield Works Engineer; W. L. Hitchens, chairman of Cammell Laird & Company; and C. E. Ellis, manufacturer, Nottingham. The Offices are at 3 Abchurch Yard, London E. C., a building which is now occupied by the head offices of Dick, Kerr & Company. 'It is hardly necessary to point out that great developments are indicated by the above facts'.

Sir Charles Ellis would emerge as the first Chairman of the Board of the new company.

Quote, Who was Who –

Ellis, Sir Charles (Edward), G.B.E., 1919; K.C.B., 1917; J.P., Nottingham; born 12th Sept. 1852; educated Lancing College; Trinity College, Cambridge, 2nd class Historical Tripos, 1875; M.A.; Barrister, inner Temple, 1878; Director John Brown & Co., Ltd., of Sheffield and Clydebank, 1890; Joined Ministry of Munitions, July 1915; late Director-General of Ordnance Supply and Member of Council of Ministry of Munitions; late Member of Royal Commission on awards to Inventors; Hon. Vice-President Institution of Naval Architects; Order of St. Anne, Russia, 1st class, Grand Officier of the Order of the Crown of Italy, Commandeur of Legion of Honneur, Commandeur of the Order of the Crown of Belgium, Osmanieh, 2nd class. Publications, Election Law and Railway Rates and Traffic, joint author. Clubs, Athenæum, Oxford and Cambridge. He died on 19th Feb. 1937.

The English Electric Co., Ltd. was incorporated on the 14th December, 1918.

The Times for Monday 17th February 1919, reported the Offer to Dick, Kerr's Shareholders –

> It is announced that holders of about 90 per cent of the shares of Messrs. Dick, Kerr, and Co. have accepted the offer to exchange their shares for shares in the English Electric Company. The latter company will now control the Coventry Ordnance Works, Messrs. Dick, Kerr, and Co., Phoenix Dynamo Manufacturing Company, the United Electric Car Company, and Messrs. Willans and Robinson. Its capital is £5,000,000 of which £1,500,000 are in Preference and £3,500,000 in Ordinary shares of £1 each. The issued capital is nearly £2,000,000.

A further report of 15th March in the Electric Railway Journal –

> It is the intention to interfere as little as possible with the internal organisation of the respective works, but the company will take over the whole of the external business management, and with certain exceptions, such as the civil engineering and contracting department of Dick, Kerr, and certain special parts of the Coventry Works, will conduct all selling and other negotiations with the public.

Kilmarnock Engineering Company:

Early in 1919 the Kilmarnock works of Dick, Kerr & Co. Limited was sold to the Kilmarnock Engineering Company and the Dick, Kerr Company became part of the English Electric Co., Ltd.

The Register of Defunct Companies has the entry –

> Dick, Kerr & Co. Ltd. registered in 1899 and went into voluntary liquidation on 21st October 1919. Those shareholders, who did not accept exchange into an equal amount of similar shares in the English Electric Co. Ltd., were entitled to 20s per preference share of £1 or 30s per ordinary share of £1. The debentures were to be repaid.
>
> The final meeting return for Dick, Kerr & Co. Ltd. was registered on 2nd July 1927.

EPILOGUE

In 1853, at the age of 23, William Bruce Dick had made the bold decision to found a new business in Glasgow based on the refining of lubricating oils, initially from vegetable and animal origin then from mineral sources. Being a qualified accountant this would appear to be well outside his sphere of knowledge or experience and requiring considerable specialist skills; after all, this was a relatively new industry involving pioneering innovation. The circumstances that initiated this move are unknown, it was, on the face of it, an ambitious project for one so young, nevertheless the decision was made and a new business was launched.

By 1862 it is clear that the oil refining business is thriving, with branches established in London and Liverpool, and Dick is seeking further opportunities for diversification into other areas of manufacturing. He had already acquired some experience of the engineering sector when in 1859 he had become the sole proprietor of the existing family business of rivet and bolt manufacturing. He identifies a need for improved fire fighting equipment in both the domestic and industrial markets and starts manufacturing fire extinguishers, invented and patented by Monsieur Vignon.

In 1875 his nephews James and John Kerr have already established themselves, independently, in the business of Iron Merchants and suppliers of railway and colliery plant, and are sharing premises, in Glasgow, with W. B. Dick. John Kerr, a qualified engineer, had already devised alternative schemes for securing rails to roadways and Dick sought to promote his skill and inventiveness by investing in the rolling stock business and promoting John Kerr to the position of Manager. At about this time, and in order to expand this part of the business and promote the Kerr systems of track laying Dick invested in the construction of the Alford and Sutton Steam Tramway soon followed by the Ipswich Tramway and the South Staffordshire and Birmingham Steam Tramway. These developments firmly established W. B. Dick in the business of tramway construction which would be diverting his time, energy and capital investment away from his core business of oil refining, which itself was continuing to expand.

When the Britannia Engineering Works became available in 1884 it was the opportunity to reduce the pressure brought about by this conflict of interests, whilst expanding the rolling stock business under the management of John Kerr. A new company was formed, Dick, Kerr and Co., handing the reins to John Kerr and promoting him to the position of managing director. From 1884 Dick, Kerr & Co., with head office in London, would adopt a totally independent business strategy; the two companies, W. B. Dick and Co., and Dick, Kerr and Co. continuing to develop along separate paths with no evidence of any business connection other than the name of Dick being perpetuated in the name of both companies and the presence of John Kerr, in 1896, on the board of W. B. Dick and Co. Ltd. William Bruce Dick continued to take an interest in the fortunes of the Dick, Kerr Company, at least from the sidelines, as indicated by his attendance at shareholder meetings and other significant milestone events to within a few years of his death.

In 1885 John Kerr is joined by the 30 year old George Flett who would prove to be a major influence on the further expansion and future developments of Dick, Kerr and Co. Before his untimely death, George Flett brought not only his obvious business talent but also connections through his other directorships in getting financial backing, most noticeably in the formation of the United Electric Car Co., Ltd. and the English Electric Manufacturing Company.

Of the other contributors to the further development of Dick, Kerr & Co., George Richardson brought his invaluable knowledge of tramway operations with the needs of the rapidly expanding tramway network and his many contacts throughout the industry and in local authorities. The Manchester Group of Messrs. Barningham, Boddington, Prestwich and Turnbull had together provided the bulk of the capital needed to establish, the Electric Railway & Tramway Carriage Works. Then in 1899 the Equipment Syndicate comprising, along with Dick, Kerr & Co. Ltd, Messrs. Kerr, Flett, Cayley, Sykes, Turnbull, R. Prestwich, J. Prestwich, Readman, Kirk and Short, provided the bulk, if not all the capital and engineering knowledge that was required to found the English Electric Manufacturing Co Ltd., some of whom were also directors of tramway operating companies and must therefore have had a considerable influence on the purchasing policy of those companies with their choice of supplier for tramway equipment and tramcars. Lastly, but not least, those talented engineers, recruited largely from the United States, who brought with them their skills and patents thereby contributing significantly to the technical and commercial success of the company.

With the formation of the English Electric Manufacturing company the Equipment Syndicate, under the leadership of Dick, Kerr & Company Ltd., brought the finest talent from the United States and 'built a modern factory on American lines'.

When competition was at its strongest the strategy employed by Dick, Kerr & Co. placed them in a unique position compared with other tramcar manufacturers in that they were able to deliver the complete package of tramway construction, the supply of electrical generation and distribution equipment and building of tramcars. The task of the customer was made simple in only needing to talk to one supplier to get a quality product at a competitive price. Dick, Kerr and their business associates were sufficiently far-seeing to combine their interests and invest in manufacturing capacity to cope with the growing demand for products that had been largely imported; this combined with an aggressive marketing strategy and an effective agency network, at home and abroad, enabled them to capture a significant proportion of a growing market. In the decade leading up to the Great War 'they set the pattern for later developments by creating a large integrated business structure in an attempt to compete with foreign conglomerates'.

An independent analysis concluded that the activities of Dick, Kerr and their associates laid the basis for the 'Rationalisation Movement' of the inter-war years. The close association of interests within the electric traction industry was used by the senior executives of Dick, Kerr to exploit a gap in British productive capacity and create a highly integrated group of companies which came to dominate the electric traction industry. Just as important, the strategy of the group and organisational developments display a high degree of sophistication in the extension of both their manufacturing and marketing.

When Dick, Kerr and Co. Ltd. became part of the newly formed English Electric Co. in 1919 the new company got the recently acquired premises of what had previously been the United Electric Car Co. on the east side of Strand Road and those of the former English Electric Manufacturing Co. Ltd. on the west side. The electrical machinery business occupying the west works and the business alliances that came with it would make it the perfect partner for the other companies with which it would soon be associated, the new company comprising –

- Dick, Kerr and Co. Ltd. of Preston, which owned:
  - United Electric Car Co. Ltd.
  - Willans and Robinson of Rugby
- Coventry Ordnance Works
- Phoenix Dynamo Manufacturing Co. of Bradford
- Siemens Brothers Dynamo Works, Stafford.

W. B. Dick & Co., the progenitor of the Dick, Kerr Company, continued to operate as a successful concern until 1945 when they were taken over by C. C. Wakefield and Co., 92 years after the company was first founded, in 1853.

The aircraft department of Phoenix Dynamo and Dick, Kerr, set up during the First World War, closed after the last English Electric Kingston flying boat was built.

Thus, came to an end the fortunes of a very successful and, for its time, pioneering company with origins in the oil refining business. By the skilful use of financial resources and the entrepreneurial talents of the founder, William Bruce Dick, and the engineering skills and business acumen of his nephew and *protégé* John Kerr, had successfully exploited new business opportunities. Perhaps, in some respects, they were fortunate in being in the right place at the right time; nevertheless they seem to have employed their considerable talents to good effect and with that typical Scottish flair for business created two successful companies, each a leader in its own sphere of activity.

Timeline

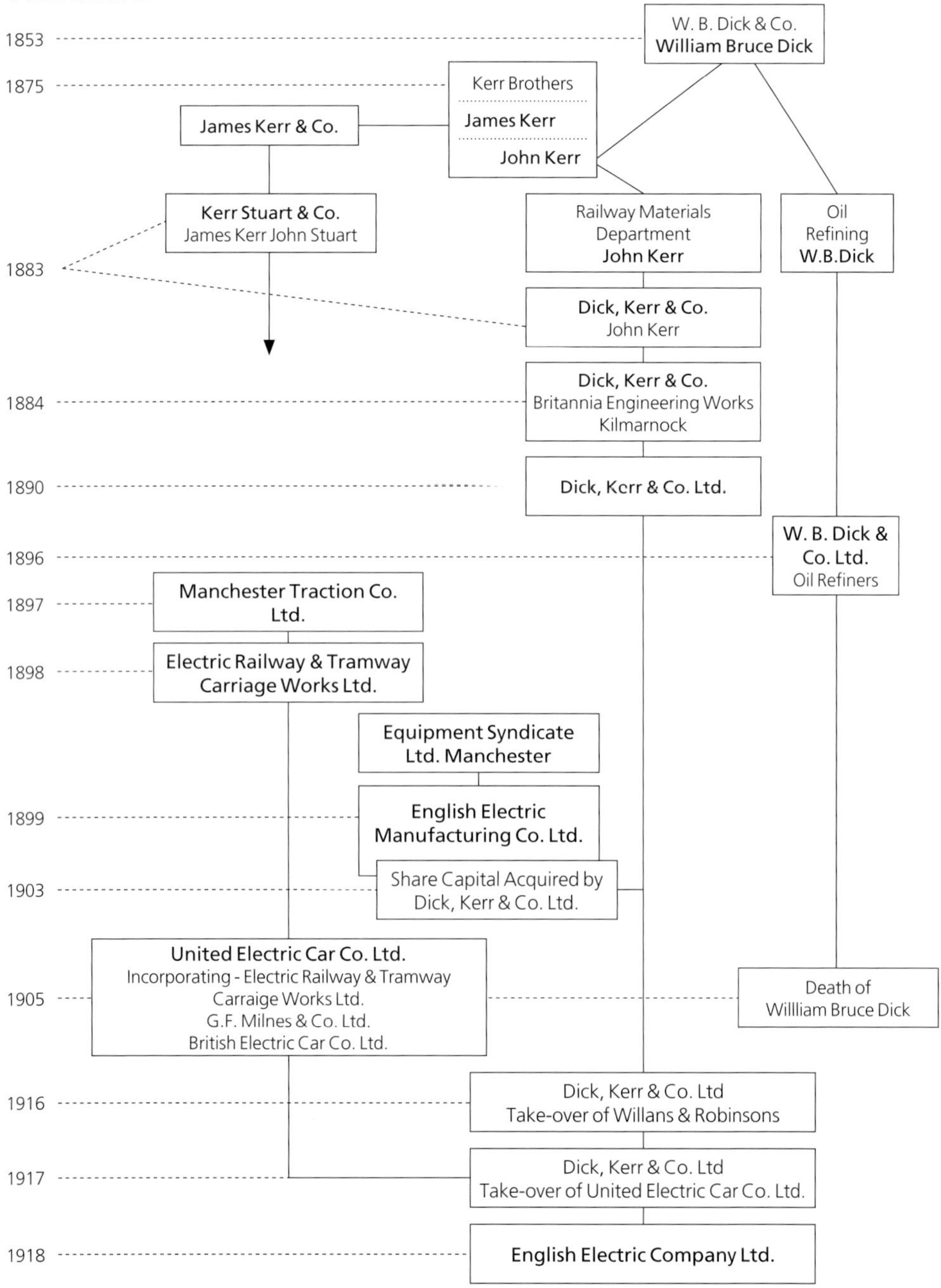
1853
W. B. Dick & Co.
William Bruce Dick
1875
Kerr Brothers
James Kerr
John Kerr
James Kerr & Co.
Kerr Stuart & Co.
James Kerr John Stuart
Railway Materials Department
John Kerr
Oil Refining
W.B.Dick
1883
Dick, Kerr & Co.
John Kerr
1884
Dick, Kerr & Co.
Britannia Engineering Works
Kilmarnock
1890
Dick, Kerr & Co. Ltd.
1896
W. B. Dick & Co. Ltd.
Oil Refiners
1897
Manchester Traction Co. Ltd.
1898
Electric Railway & Tramway Carriage Works Ltd.
Equipment Syndicate Ltd. Manchester
1899
English Electric Manufacturing Co. Ltd.
1903
Share Capital Acquired by Dick, Kerr & Co. Ltd.
1905
United Electric Car Co. Ltd.
Incorporating - Electric Railway & Tramway Carraige Works Ltd.
G.F. Milnes & Co. Ltd.
British Electric Car Co. Ltd.
Death of Willliam Bruce Dick
1916
Dick, Kerr & Co. Ltd
Take-over of Willans & Robinsons
1917
Dick, Kerr & Co. Ltd
Take-over of United Electric Car Co. Ltd.
1918
English Electric Company Ltd.

BIOGRAPHICAL NOTES

The Founders
William Bruce Dick

William Bruce Dick was born on 23rd May 1830 at Maybole, Ayrshire the third child of William Dick a Merchant and his wife Mary Bruce. His siblings were Mary Dick, born in 1827; Margaret Bruce Dick, born in 1828 (*mother of John Kerr*), and Catherine, born in 1831. William Bruce Dick was married in 1861 to Elizabeth H. Harker, they had ten children, the first eight born in Glasgow, the last two in Putney.

In the census return for 1871 William B. Dick is an Oil Refiner aged 40 with his family and living in Gardener Terrace, Cecil Street, Govan, included in the household are their first six children and five servants.

In 1881 William and Elizabeth with nine children, Elizabeth's sister and six servants are at Carlton House, Putney Hill, London; William is a Merchant aged 50 and born in Scotland. In the same census their eldest son (William) Bruce Dick is unmarried and resident in Bath Street, Barony aged 18 and an Oil Merchant and Refiner.

In the 1901 census William, aged 70, is resident at the Central Station Hotel, Glasgow, his occupation is Chairman of W. B. Dick.

A newspaper interview with Dick In 1903 provides us with an interesting biographical profile which reveals something of his personality and peripheral business interests towards the end of his life –

In 1879 William Bruce Dick had acquired the Spa Hotel at Tunbridge Wells, as a pure speculation, and proceeded to enlarge and improve the buildings and facilities considerably. He bought all the available land contiguous to the holding and eventually owned 65 acres. The Spa was a first class residential hotel, its grand dining room measured 60 ft by 40 ft; the principal drawing room was 63 ft by 40 ft, there was a theatre with a large stage and a separate concert room. The bathing facilities included shower, Russian, Turkish, medicated and a swimming pool measuring 40 ft by 15 ft. Massage was practiced and electricity in the form of a galvanic current could be applied to patients or visitors, who require it, at moderate charges (no pun intended); the facilities also included a post office and golf course. The Hotel generated its own electric lighting supplying the 520 lights from a dynamo driven from a 25 hp engine. The contingency of fire occurring on the premises was provided for by efficient fire extinguishers (*appropriate since Dick held patents for a range of fire extinguishers manufactured, in his early business career, by W. B. Dick & Co., then in later years under licence by the Fire Appliance Co. and which contributed considerably to his wealth*). The newspaper correspondent in assessing the character of W. B. Dick describes him as a person 'who appreciates the beauties of Nature, has a poetic fancy for meandering streams, fairy dells, rustic bridges and lakes adorned with lilies and swans – hence his development of these features in his grounds where he loves to sit on a rustic seat, in a shady dell, smoking a good cigar, and contemplates the varied and picturesque scenery which he loves so well'.

In addition to the property at Tunbridge Wells W. B. Dick owned a beautiful property called Carrick Grange near Sevenoaks, and had moved there from Putney Hill, London where he had lived since 1875. He also rented a farm 'Westwood' of 205 acres lying between Sevenoaks and Tunbridge Wells. 'Being a canny Scot with a well developed sense for business he believed in making his hobbies pay'. --- Mr. Dick indulges himself by motoring in his 24 horse power Darracq automobile between Carrick Grange, Westwood and Tunbridge Wells, with occasional visits to town and his native heath, bonnie Scotland.

The golf course that Dick created in the grounds of the Spa Hotel had originally been a modest three or four-hole course, constructed for his own amusement and that of special guests. This was later extended to a nine-hole course, becoming the home of the Tunbridge Wells Golf Club founded by William Bruce Dick and first registered at the Royal and Ancient Golf Club of St. Andrew's in 1896.

William Bruce Dick died on 11th July 1905, aged 75, at Carrick Grange, Sevenoaks, Kent.

In his will, probate granted 14th September 1905, to William Bruce Dick, Merchant, George Lucas Pardington, M.D., Frederick Richard Leftwich, Gentleman and John Ebenezer Ranking M.D., effects £233,349 -2s-5d, double probate March 1906.

His obituary, published on 11th July 1905 –

Mr. William Bruce Dick founder of the electrical engineering firm of Messrs. Dick, Kerr and Co., died suddenly this morning at the age of 75, at Carrick Grange, Sevenoaks.

(*It is interesting that Dick, Kerr and Co., is given prominence with no mention of W. B. Dick and Co.*).

William Bruce Dick and Charles E. Dick sons of William Bruce Dick had been Directors of W. B. Dick and Company from its incorporation in 1896. On the death of their father William Bruce Dick, the elder brother, became Chairman and Managing Director and Charles E. Dick became joint Managing Director, they had been managing the company since 1896.

In the census return for 1911 William Bruce Dick aged 48 a Director of Companies has been married for 13 years and is living in Crawstone Road, Westcliff, Elmfield, Essex with his wife Alberta Mary, two children Sybil Mary Margaret Bruce Dick and Owen Robert Bruce Dick and four servants.

Charles E. Dick would appear to have had more than a little talent for the game of golf. In the history of the Dornoch golf club we find in 1901 Mr. Andrew Carnegie presenting a 'splendid silver shield' for open competition at Dornoch. Right from the start the August meeting with the shield as the main trophy, drew prominent golfers from far and wide to Dornoch links. Notable golfing names have been inscribed upon the bosses of its silvery surface: Ernest Holderness, Charles E. Dick, Roger Wethered, J. Gordon Simpson, as well as local Dornoch names".

John Kerr

In 1850 Margaret Bruce Dick, the fore-mentioned sister of William Bruce Dick married John Kerr, a Manufacturer of Silk, Wool and Cotton, at Barony, Lanarkshire. They had four children, James Kerr, born C1851; John Kerr, born 22nd August 1852; Mary Bruce Dick Kerr, born in 1854 and Margaret S. Kerr, born in 1856 all at Barony, Lanarkshire.

The 1851 census for Scotland gives John Kerr, aged 30, a Manufacturer (Silk Wool & Cotton) one of 2 employing about 600 persons, and living at 3, Lyndock Street, Barony; with him are his wife Margaret, aged 22; Sister in Law, Catherine Dick, aged 18, both born at Maybole and two servants, a Housemaid and a Cook.

In 1861 James Kerr, aged 9 and John Kerr, aged 7 are at a Boarding School at 2 Cassilis Street, Glasgow.

In the 1871 census Mrs John Kerr (*Margaret*) by now a Widow aged 41 with family, James, aged 19; John, aged 18; Mary B. D. (B*ruce Dick*), aged 17 and Margaret S., aged 15 is at 5 Viewfield Terrace, Ann Street, Glasgow; Margaret is an Art Student at Glasgow University. With them is Catherine Brodie (*formerly Dick*), her Sister, and four servants.

John Kerr married Helen Margaret Melville, of Falkirk, in about 1877. At the time of the census John Kerr, aged 28, an Iron Merchant is living at Oxford Road Mayfield, Putney, London with Wife Helen, children John Bruce Kerr, aged 2 and Helen M. M. Kerr and three servants.

By 1901 John Kerr has made a new home for his family, which seems to reflect his elevated status as a man of wealth with a growing industrial empire. His address is given as Gilmerton House, Athelstaneford, Haddingtonshire, East Lothian. John and Helen with four daughters have a staff of eleven servants including a Butler and two Footmen to care for their needs.

Ref: Who Was Who 1916–1928:-

Kerr, John b1852; m1877 Helen Margaret d of Thomas Melville of Kersehill, Stirlingshire; two s, four d, M.P. (U) Preston, 1903 – 1906; address: Bramshill, Frinton-on-Sea, Essex.

The Politician:

The General Election of 1900 was John Kerr's first attempt to break into national politics. Reported in the Pall Mall Gazette on 13th October 1899 –

> Mr. John Kerr, Gilmerton House, who has been adopted as the Unionist candidate for East Lothian at the next election, is the head of the firm of Messrs. Dick, Kerr, and Co., contractors for the Edinburgh tramway cabling and many other large undertakings; he failed at this attempt by a small margin although reducing the liberal majority.

In April 1903 Robert William Hanbury, the Conservative MP for Preston, died aged 58. John Kerr, having unsuccessfully contested the 1900 general election in his home area of Haddingtonshire, was chosen to contest the by-election in Preston, standing as the Unionist candidate; he won the election with a majority of 2,149.

During a short and intense period of electioneering, which seems to have started within days of the death of Mr Hanbury, the local press seem to have enthusiastically supported the candidature of John Kerr.

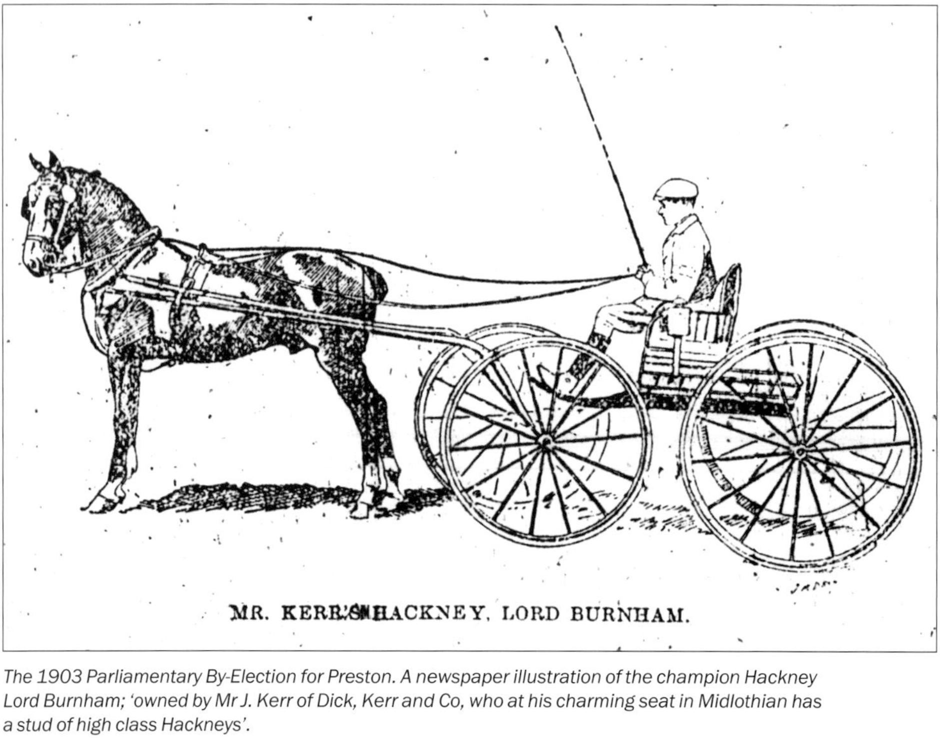

The 1903 Parliamentary By-Election for Preston. A newspaper illustration of the champion Hackney Lord Burnham; 'owned by Mr J. Kerr of Dick, Kerr and Co, who at his charming seat in Midlothian has a stud of high class Hackneys'.

The Lancashire Daily Post of 8th May 1903 describes John Kerr as –

> A thorough sportsman and breeder of pedigree Hackneys; having taken many first prizes in single, double and tandem at the leading agricultural shows. He has a Scotchman's love of football, and used to play both over the border and in London. He is keen on golf, fishing, shooting and all those recreations which have contributed to the development of a form almost as massive as that of our late member. Mr Kerr fought Haddingtonshire at the last election, the voting being - Haldane K.C., 2668; Kerr, 2290 – majority 378. The fight was a great one, and though Mr Kerr did not succeed in winning, he pulled the Liberal majority down handsomely. Personally, Mr Kerr is a gentleman of an unassertive John Bull type, who means to make a good fight.

On 12th May a caricature of John Kerr has the caption 'He wanted them to understand that in Scotland he was known as a Presbyterian and when he was in this country he was an Episcopalian'. On 13th May a further caricature with a quote from his opponent, Mr Hodge "Mr Kerr has been beating the drum of British Imperialism". The result of the election reported on 16th May was Kerr (U) 8,639; Hodge (Lab) 6,490; majority 2,149. The Guardian for 16th May, commenting on the election has a caricature of Mr Kerr's Hackney, 'Lord Burnham'.

The 1903 Parliamentary By – Election for Preston. John Kerr Beating the Big Drum.

The 1903 Parliamentary By – Election for Preston. John Kerr Taking the Broad View. "In Scotland he was known as a Presbyterian and when he was in this country he was an Episcopalian".

On 12th June 1903 there is a report on **Mr. Kerr's Expenses** –

> A return of the expenses of Mr. John Kerr, the Conservative candidate who successfully contested the recent by-election at Preston, has just been issued The total expenditure was £1,044 2s, made up as follows: Returning officers charges, £192 9s 7d; Agents, clerks, messengers and polling agents, £369 12s 2d; printing, stationery, advertising and bill-posting, £268 0s 10d; postages, hire of rooms, and other petty expenses, £120 3s 11d; and Mr. Kerr's personal expenses, £23 15s 6d; total £1044 2s 0d.

In 1903 John Kerr acquired the lease on Loudwater House, a large property near Rickmansworth in Hertfordshire. The property came with a purpose built carriage house to accommodate the horses and hackney carriages with which he competed successfully at shows throughout the country.

His many successes included –

In 1905, and the second day of the Royal Agricultural Society's show at Park Royal attended by the King and Queen. The day's judging was confined to double harness horses for which John Kerr, M.P., Gaddesden Place, Hemel Hempstead with "Advertisement" and "Paddock Wildfire" was awarded a first place and silver cup.

At the International Horse Show at Olympia in 1907 he received awards for show horses "Loudwater Little John" and "Loudwater Robin" and in a tandem class a blue ribbon with "Loudwater Little John" and "Loudwater Robin Hood".

In 1909 at the Richmond Horse Show he was awarded the cup for the champion harness horse for the second time in succession with "Loudwater Flourish" becoming the owner of the trophy.

In 1910 he was elected President of the Hertfordshire Agricultural Society for the ensuing year.

He also appears to have enjoyed the occasional day out on the moors grouse shooting – In August 1905 a party from Cowshill Hotel in Weardale, which included Mr John Kerr M.P., Messrs. James Woolston, Barker, Woodroffe, Piper, Wilkins, Robertson and Kennedy, found "splendid sport" on the Irishope moors, and heavy bags resulted.

John Kerr has a single entry in Hansard, for 15th June 1904 –

Mr Kerr (Preston) formally moved under section 9 of the Private Legislation Procedure (Scotland) Act, 1899, that the Leith Corporation Tramways Order Confirmation Bill be referred to a Joint Committee of Lords and Commons.

This was seconded by Sir Lewis McIver (Edinburgh, W.) with the opening comment "as his hon. Friend who had moved this motion, being pecuniarily interested in this matter, could neither speak nor take part in the division upon this Question, the duty of explaining the reason for this motion being made devolved upon him".

The run up to the General Election of 1906 must have been a painful experience for John Kerr. Uproarious scenes were witnessed at Conservative gatherings at Preston last night, the retiring members, Sir William Tomlinson and Mr. John Kerr, being subjected to a running fire of interruptions and hooting, and the proposed vote of confidence being lost, the proportion of hands held up against it being six to one.

A newspaper leader of 23rd November 1906 – Mr. Kerr's Test.

> "Appropriate Melancholy" at a Tory Meeting, Mr. John Kerr, ex M.P., presiding. An appropriately melancholy note was struck by the chairman in his opening remarks, for he said: The result of the last election was painful to them --- Mr. Halsey fell, like other good men. Mr. Marshall Hall fell with his flag flying, and he himself went under. That was nothing. He spoke more in sorrow than in anger. He deplored the extreme ignorance of the electors.

The leader writer goes on to say –

> Mr. Kerr follows the old plan of making agreement with him-self the test of intelligence, and then naturally enough, he discovers that his opponents are deplorably ignorant! The good man has apparently not been endowed with a sense of humour. It must have been an impressive moment when Mr. Kerr mourned over the ignorance of his countrymen, not in anger but in sorrow, and the other two victims of the same misfortune sat by, wagging their pates in sad sympathy.

The Times for Saturday 1st January 1910, reports

London Bank of Mexico and South America, Mr John Kerr chairman of Dick, Kerr & Co. has joined the board.

The entry for John Kerr in Who's Who of British MPs gives the following information –

> Kerr John, Gaddesden Place, Hemel Hempstead, Herts., Junior Carlton, S of John Kerr Esq. B 1852, Educated at St. Andrews and Glasgow Universities, Engineer, Chairman of Dick, Kerr & Company Ltd., of Kilmarnock and Preston. A J.P. for East Lothian, a conservative, unsuccessfully contested Haddingtonshire in 1900, elected for Preston May 1903 and sat until defeated in 1906.

John Kerr retired from the Board of Directors of Dick, Kerr and Co. Limited in October 1912, after 37 years in the business and 29 years after the founding of Dick, Kerr and Co. His lease on Loudwater House was terminated in 1913 and he moved to Weardale Place, St. John's Chapel, County Durham. He appears to have had financial problems, for on 11th December 1913 he is declared bankrupt.

The Times newspaper on Monday 9th March 1914 has an advertisement for a forthcoming auction to be held on 17th and 18th March -
John Kerr – in Bankruptcy; Loudwater House, Rickmansworth; the contents of the residence, for auction, included a 28-36 Daimler Landaulette, three punts, garden tools, dog kennels, barns, loose boxes and various items of furniture. He subsequently moved to Tendring, near Frinton on Sea, Essex where he died in March 1925, aged 72.

James Kerr

James Kerr was born in Glasgow on 14th April 1851; he went on to study arts at the University. On 7th February 1874, James Kerr was elected a member of the London Association of Foremen, Engineers and Draughtsmen at the ordinary monthly meeting of the Association held at the City Terminus Hotel. James Kerr died on 4th December 1884 at the premature age of 33.

The Financiers
Robert Slater Boddington

Boddington was one of a small group of Manchester based industrialists who contributed to the financing of the tramcar manufacturing business on Strand Road, sometimes styled the Manchester Group. He served on the boards of the Blackpool and Fleetwood Tramroad Company, the Electric Railway & Tramway Carriage Works and the United Electric Car Co. Ltd.

Robert Slater Boddington was chairman of Boddingtons Brewery, Strangeways, Manchester, from 1908 following the death of his brother and former chairman William Slater Boddington. He was born on the 14th October 1862 in Eccles, baptised at Christ Church, Patricroft and died on 1st September 1930, aged 67, at Kerfield, Knutsford. His parents were Henry Boddington (1813-1886), who had first joined the brewery in 1832 as a travelling salesman, and Martha Slater (1819-1871). Robert married Maud Crudelius (1863-1941), daughter of Rudolph Wilhelm Crudelius and Mary McLean of Leith, on 3rd February 1886, they had four sons. In 1881 he is resident in Lytham, in 1891 in Higher Broughton, Salford and in 1901 and 1911 in St. Annes on Sea. His entry in the national probate calendar states Robert Slater Boddington of 35 New Bridge-street, Manchester and of Kerfield, Ollerton near Knutsford, Cheshire died on 1st September 1930 at Kerfield; probate London 19th December to Frank Boddington fruit grower, Alan Boddington stockbroker and Philip Boddington and Charles Geoffrey Boddington brewers, effects £94,392 6s 5d.

Strangeways Brewery had been founded in 1778 by two grain merchants, Thomas Caister and Thomas Fry, their principal customers being the cotton workers of Manchester. Henry Boddington, born in 1813 in Thame, Oxfordshire, joined the Brewery in 1832 as a travelling salesman. By 1848 he had become a partner and in

January 1853 he borrowed money to become its sole owner. Henry had been born in a mill cottage at Thame (a blue plaque marks the event) where his father, John Boddington, was the Miller. After Henry Boddingtons's death, his son, William Slater Boddington became company chairman, and the company went public in 1888 with assets of £320,465, and was known as Boddingtons Breweries Ltd. After the death of William Slater Boddington in 1908 his brother, Henry's youngest son, Robert Slater Boddington, took up the reins and completed a 50 year association with the company until his death in 1930.

Robert Breese Barningham-

A director of the United Electric Car Company then, in 1914, its chairman, he was a director of the Manchester Traction Company, the City of Carlisle Tramways and the Blackpool and Fleetwood Tramroad Company.

An iron and steel merchant, he was born on 25th January 1850 in Manchester the son of Thomas Barningham (1810-1871) and Sarah Breese (1814-1880) and died, age 81, on 26th January 1932 in Ealing, Middlesex, with estate valued at £30,091-3-4. He married Annie Elizabeth Worthington in about 1880, they had three daughters. In 1881 he is living in Stretford and in 1901 in Withington.

Kelly's Directory for Manchester in 1895 gives Robert B. Barningham of Barningham Brothers, Saxonholme, Upper Chorlton Road, Coal and Iron Merchant.

Slater's Directory for Manchester in 1905 has the entry for Barningham Brothers, Iron, steel, coal and coke merchants of 41, Victoria Buildings, Victoria Street and Midland Coal Wharf, Ashton Old Road.

- James – Coal & Iron Merchant (Barningham Brothers), Beech Mount, Dunham Massey.
- Robert B. – Coal & Iron Merchant (Barningham Brothers), Clough Fold, Wood Road.
- Thomas – Householder, 58 Eccles Old Road.

Kelly's Directory for 1909 has Barningham Brothers, Iron, steel, coal and coke merchants of 41, Victoria Buildings, Victoria Street and L & Y Railway Yard, Hope Street.

- James – Coal & Iron Merchant, Beech Mount, Dunham Massey.
- Thomas – Coal & Iron Merchant, The Limes, Eccles Old Road.

The London Gazette on 5th April 1927 gives notice that, the old established business of Steel, Iron and Coke Merchant, carried on for many years by the undersigned Robert Breese Barningham, formerly in partnership with his late brother, James Barningham and latterly alone under the firm name of Barningham Brothers, at 41 Victoria Buildings, Manchester, has been disposed of by him as from 31st March, 1927, to the undersigned, Oliver Greenwood, who will continue to carry on the business at the same address and under the same firm name.

Frank Dudley Docker, C.B.

Frank Dudley Docker was born in Smethwick on 26th August 1862 and died on 8th July 1944, the youngest son of Ralph Docker. In 1881 he joined in forming Docker Brothers, with his elder brother Ludford Charles Docker, dealers of varnish for blacking stoves and later manufacturing their own varnishes, for railway carriages and wagons at home and abroad. It was as manufacturers of varnish that they are found in the accounts of the Electric Railway and Tramway Carriage Works. In 1895 he marries Lucy Constance Hebbert and has one child Bernard (1896-1978), later Sir Bernard Docker. For recreation he played golf and he was a member of the Carlton Club.

In 1902 Dudley Docker masterminded the amalgamation of several railway carriage makers, to form the Metropolitan Amalgamated Railway Carriage and Wagon Company, Limited, of which he became chairman. The company was registered on 18th April as a limited company, with works and company offices, at Saltley, Birmingham, to amalgamate:

- Ashbury Railway Carriage and Iron Co.
- Brown Marshalls and Co.
- Lancaster Railway Carriage and Wagon Co.
- Metropolitan Railway Carriage and Wagon Co.
- Oldbury Railway Carriage and Wagon Co.
- Willingsworth Iron Co. Ltd.

The Patent Shaft and Axletree Co. were acquired later that year and in 1907 Docker Brothers were acquired.

Docker was an advocate of industrial rationalisation through merger, and Birmingham's 'industrial titan', who had risen to national prominence through effecting the amalgamation of the five railway carriage and wagon companies to form the MARCWC in 1902. Described as 'a showpiece of modern, rationalised combination', the MARCWC became the largest and most powerful rolling-stock manufacturers in Britain, eventually integrating production, distribution and marketing. The MARCWC also collaborated with Dick, Kerr & Co. Ltd. that under Docker's initiative became a central player amongst a group of firms who responded to the foreign challenge in the field of electrical engineering. Such was Docker's reputation 'for business infallibility' that he joined the boards of several companies prior to the war, where he played a leading role in 'arranging mergers'; one of these companies was BSA, which Docker joined early in 1906.

In 1906 he is a Director of Birmingham Small Arms (BSA) and in 1910 masterminded the purchase by BSA of Daimler Motors. Docker's influence on boardroom selection at BSA is illustrated by the appointment of Lincoln Chandler, his colleague on the board of the MARCWC, as a BSA director in 1907.

Docker's association with George Flett, who sat on the board of the MARCWC and Daimler and was a leading executive at another Docker interest, Dick, Kerr & Co. may well have aided further his knowledge of Daimler's business affairs.

Docker's arrival at BSA also coincided with the decision to create a secret reserve, a practice pursued by the MARCWC. On its formation in 1902, the articles of association empowered the directors of the MARCWC to build up secret reserves not published in the balance sheet; by 1910 these amounted to £450,000. Apparently, the MARCWC used reserves for dividend smoothing, especially in the depressed year of 1909, but also fulfilled the ambition of Docker for funding potential future acquisitions. In 1904 Docker in association with John Kerr, George Flett and William Samuel Laycock financed the acquisition of G. F. Milnes & Co. and the British Electric Car Works and their merger with the Electric Railway & Tramway Carriage Works to found the United Electric Car Co., Ltd. In 1914 Docker bought a London evening newspaper, *The Globe*, to agitate for greater influence of industrialists in government policy towards business.

By 1914 the MARCWC had about 14,000 employees and occupied 475 acres of factory space. They would become one of the UK's key manufacturers of guns, armoured vehicles and finally tanks for the Great War effort, and ultimately made about 80 percent of all tanks deployed by Great Britain during the war.

In 1912 the name of the company was changed from Metropolitan Amalgamated Railway Carriage and Wagon Company, Ltd, to Metropolitan Carriage Wagon and Finance Co. Ltd. Dudley Docker was also a Director of the Midland bank (1912), the Electric and Railway Finance Corporation and the Metropolitan Railway (1915) a founder of Midlands Employers Federation (1913),and in 1916 was founder President of the Federation of British Industries. At the political level Docker was President of the British Commonwealth Union (BCU), a right wing pressure group. In 1917 he was instrumental in arranging the purchase by Metropolitan Carriage Wagon and Finance Co. Ltd. and Vickers of British Westinghouse then persuaded Vickers to buy Metropolitan Carriage Wagon and Finance Co. Ltd. at 'an inflated price' to form Metropolitan-Vickers Electrical Company. Frank Dudley Docker was a Director of Vickers until 1920.

Frank Dudley Docker was one of the three most significant sponsors of Shackleton's Endurance expedition to the South Pole, together with Sir James Caird and Dame Janet Stancomb-Wills. He gave £10,000 towards the purchase and refit of *Endurance*; and had one of the ship's lifeboats named after him, by Shackleton, the 'Dudley Docker'. The 'Dudley Docker' remained on Elephant Island with the 'Stancomb Wills' and formed a shelter for the marooned members of the expedition whilst the 'James Caird' took Shackleton and its crew to South Georgia.

In the 1880's Docker played several games of county cricket for the Midland counties. He served on the committee of the newly founded Warwickshire County Cricket Club until 1892, having been the highest scoring batsman at the inaugural match of the Edgbaston ground in June 1886 (when he was voted man of the match); he also played for his county against Australia the same summer, and against the 'Gentlemen of Canada' the following year. He was a keen shot, and spent time in Scotland each year, where he enjoyed fast walking on the moors, and was a member of the Royal Thames Yacht Squadron. Docker was made a Commander of the Bath in 1911.

Frank Dudley Docker died on 8th July 1944 in Amersham. Bernard Docker, his only son, became Managing Director of BSA from the early 1940s and was also Chairman of Daimler until 1956.

Ludford Charles Docker, J.P.

Ludford Charles Docker, brother of Frank Dudley Docker, was Chairman of Docker Brothers and as a Director of the MARCWC he represented the interests of that company on the Board of the United Electric Car Co. Ltd. from 1905. He was born on 26th November 1860 in Smethwick, the son of Ralph Docker and Sarah Maria Docker and married, Marie Park daughter of J. C. Park, C.E. from Dublin on 28th July 1894; they had three children born in Edgbaston; he was educated at King Edward VI Grammar School, Birmingham. In the census returns for 1901 and 1911 he gives his profession as Paint and Varnish Manufacturer, presumably carrying on the family business whilst also serving as a Director on the board of the parent company.

Docker played cricket for Derbyshire from 1881 to 1886, being captain for a time, and then went to Warwickshire, for whom he appeared regularly until 1894. He became president in 1915, retiring in 1931. His highest score in county cricket was 163 at Edgbaston against Cheshire in 1891, while three years later his average, 37.20, was the best by any gentleman in first-class cricket. He visited Australia in the winter of 1887 with Shrewsbury's Cricket Team, but met with little success, his best innings being 48 in eleven-a-side matches. He played his club cricket with Handsworth Wood, in the Birmingham League and in 1886 presented the Docker Shield for competition between Birmingham Schools. He was High Sheriff of Warwickshire 1923-24, was a member of the Conservative Club in Birmingham and for recreation he played golf.

Thomas Dewhurst Lingard

Thomas Dewhurst Lingard was a director of the English Electric Manufacturing Co. Ltd (from 1900) and a director of the City of Carlisle Tramway Co.

He was born in 1848 in Birkenhead Cheshire to Alex R. Lingard, a Surgeon, from Manchester and his wife Sarah; he was married in 1883 to Regina Caroline Walpole at Newport Pagnell in Buckinghamshire. In 1881 Thomas D. Lingard a Solicitor, aged

33, is living at 3 Lime Grove, Chorlton on Medlock with his parents and six servants. In 1891 he is resident at Burnside, Windermere, a Solicitor, with wife Regina, a son John, a Governess and five servants, and, in 1911 at Windermere, a Solicitor aged 63, with wife Regina, a visitor and four servants.

Information concerning his family crest reveals –

Esquire of Burnside, Windermere, a wolf's head erased sa., charged on the neck with an escallop arg., and holding in the mouth a branch of three cinquefoils slipped vert. Born Fellside, Windermere, died 14th January 1925, married Caroline Walpole. *The place of birth (Windermere) is incorrect.*

William Samuel Laycock, M.I.M.E.

Born in Sheffield on 20th October 1842, William Samuel Laycock was the son of W. E. Laycock, Mayor of Sheffield in 1865 and a member of the firm of Samuel Laycock and Sons, hair seating manufacturers of Sheffield. He was educated at the Collegiate School, Sheffield. It has been claimed that 'he had done more than any other individual in England to promote the comfort and secure the safety of the railway passenger by his many inventions for the perfecting of the equipment and conveniences of railway rolling stock'.

As a manufacturer of railway material he founded the Victoria Works, Sheffield, but these becoming too small for a rapidly expanding business he planned and erected at Millhouses, a fine modern manufactury. He regularly travelled to America and was a keen student of engineering progress at home and abroad. A Member of the Institution of Mechanical Engineers, elected 1883; Justice of the Peace for Sheffield he had made his home at Oakbrook, Fulwood Road, Sheffield.

Laycock, as a director of the British Electric Car Co. Ltd., was one of the businessmen, who in 1905, along with John Kerr and George Flett, invested their own capital in the bringing together of the Electric Railway & Tramway Carriage Works with the Castle Car Syndicate and the British Electric Car Co. to form, with additional financial backing from the Metropolitan Amalgamated Railway Carriage and Wagon Company, Ltd., the United Electric Car Co. Ltd.

In 1910 he became chairman of Craven's, Ltd., rolling-stock builders, of Darnall, Sheffield, and re-equipped the works with modern machine tools to undertake the building of Pullman cars and other types of high-class passenger rolling-stock for British and foreign railways, in addition to wagons, which had been their main product.

In the 1911 census, aged 66 and an Engineer, his occupation is that of a Government & Railway Contractor and living in Sheffield; he died at home on 2nd March 1916 at the age of seventy-four.

Frederick William Duthie Manuelle

On the formation of Dick, Kerr & Co. Ltd. in May 1890 there was an 'unregistered agreement' between F. Manuelle and John Kerr, of the first part; G. Flett of the second part; and G. W. Carr-for the company-of the third part; Messrs Manuelle and Kerr being the promoters of the newly registered company. Manuelle, along with John Kerr, was also a founding director of the Harrow Road and Paddington Tramways Company and in 1894 the Edinburgh and District Tramways Company.

Frederick Manuelle was born in Rotherhithe and baptised on 2nd November 1826 at Southwark; his parents were Charles Manuelle, a stone merchant and Mary. He married Maria Jane Core on 9th August 1851 at Islington and died on 11th April 1907. Probate records show 'Frederick William Duthie Manuelle of 100 Queens Gate, Kensington and 57 Gracechurch Street, London, died 11th April 1907 at Mentone, France; probate London 25th May to Frederick William Manuelle, merchant & Harold Brown, Solicitor – effects £116,429-4-8'.

Frederick is a stone merchant living in Rotherhithe in 1841; a granite merchant in Islington in 1861 and in Camberwell in 1871. From 1880 to 1907 he is living at 100 Queens Gate, Kensington. His business address in March 1888, given in the High Court of Justice – Chancery Division, is at 101 Leadenhall-street, also the business address for John Kerr, and, on the formation of Dick, Kerr and Co. Ltd. in 1890 at 131 Leadenhall-street; he is a granite merchant and quarry owner. In 1862 we find him contracting to supply pavement – 10,000 feet of 2½-in Tooled York Pavement for £248-0-0 and 5,000 feet of Devon or Cornish Granite Kerb for £249-15-0.

His son Frederick William Manuelle took over and must have made a considerable success of the business, for in 1898 he is living at 4 Park Lane, Westminster; in the 1901 census at Piccadilly, St. George Hanover Square, a granite merchant with wife Joanna and several servants. He died on 21st August 1916, aged 47, at 80, Bishopsgate, London, home address St. George Hanover Square; probate 19th January 1917, £79,389-5-9.

On 18th July 1911 he applied to be admitted to the Freedom of the City of London, by redemption, in the Company of Shipwrights. Frederick William Manuelle, son of Frederick William Duthie Manuelle late of 100 Queens Gate, Granite Merchant & Quarry Owner died occupying premises 57, Gracechurch Street, in the City of London, Granite Merchant & Quarry Owner.

Richard Henry Prestwich

Richard Henry Prestwich was a cotton spinner born in January 1847 at Farnworth, Bolton, and with his brother John Edward Prestwich he was partner in the Pandora Mill on Longcauseway, Farnworth, which had been founded by their grandfather.

A title deed, to premises in Farnworth, and dated 8th November 1905 names John Edward Prestwich and Richard Henry Prestwich of Pandora Mills, Long Causeway, Farnworth as holding a freehold plot of land of 3,500 and 25 square yards with 21 dwelling houses erected thereon; numbers 1-19 Mary Street and 22-42 Howarth Street and the waterworks office & store yards in Long Causeway.

The Cotton Trade Directory, for 1923, lists Prestwich (S & J) Pandora Mills, Farnworth (branch of Burberry's Ltd.) with office at 13 Marsden Square, Manchester. With others they developed a technique for weatherproofing garments and which were used by the first Arctic and Antarctic explorers. In 1920 the firm was amalgamated with Burberry's.

Richard was a director of the Electric Railway and Tramway Carriage Works, and the United Electric Car Company Ltd., then Chairman of the English Electric Manufacturing Co. Ltd. He was also a director of the Blackpool and Fleetwood Tramroad Co. His brother John was a debenture Trustee, along with Thomas Strover Turnbull of the Electric Railway and Tramway Carriage Works, and later, the United Electric Car Co. Ltd, in which both Richard and John had a significant shareholding. A trust deed exists, dated 28th November 1901, between the Electric Railway and Tramway Carriage Works Ltd., and John Edward Prestwich and Thomas Strover Turnbull for £50,000 5% first mortgage debenture stock.

Richard was a partner in the group of capitalists, styling themselves the Equipment Syndicate Ltd, with a registered office at 13 Spring Gardens Manchester, the office of Thomas Strover Turnbull, Solicitor. This group were initially responsible for establishing and building the Works on the west side of Strand Road which was very soon bought by Dick, Kerr & Co. (also a partner in the Syndicate) and set up as the English Electric Manufacturing Co. Ltd. Along with Thomas Strover Turnbull, Richard was a trustee of the company and in 1902 they are seen to be jointly holding 38,800 shares.

In 1911 Richard Henry Prestwich, aged 60, is resident at Higher Broughton, Salford, with wife Rebecca, two daughters and four servants; his occupation is that of Cotton Spinner, manufacturer.

Benjamin Corless Sykes, M. Inst. C.E.

In 1898 Benjamin Sykes was instrumental along with John Kerr and George Flett in establishing the Electric Railway and Tramway Carriage Works on the East side of Strand Road, then in 1900 he was a member of the syndicate that financed the newly established English Electric Manufacturing Co. Ltd. on the west side of the road. The existing and neglected works on the East side of Strand Road had originally belonged to the North of England Railway Carriage & Iron Co., Ltd. which had been registered in 1863. The freehold passed, in 1871, from Preston Corporation to Henry Green, Frederick Crimes, Benjamin Sykes and Thomas Whitehead, acting on behalf of the

North of England Railway Carriage & Iron Co., Ltd. The company eventually wound up in 1878 and the equipment removed by 1880, leaving the buildings as empty shells. When the Dick, Kerr Company was looking for manufacturing premises in 1898 it is probable that Benjamin Sykes acted to make the site available to the company, which would explain his financial interest and why on 7th October 1898 the board approved the payment of £20,000 to Messrs Sykes, Kerr and Flett, on account. The location would have been ideal for the new company with its easy access to the railway network and to Preston Docks.

Benjamin Sykes was a civil engineer and partner to Edward Garlick, resident engineer for the construction of Preston Dock (1884 – 1892). His obituary of August 16th 1907 –

The death occurred at his residence, Elston, near Preston, on Monday, of Mr. Benjamin Sykes, M. Inst. C.E., who engineered several prominent Lancashire schemes, in addition to being connected with the establishment of the Electric Car Co., Ltd., which employs between two and three thousand men in Preston, he carried out important work in connection with the Blackpool and Lytham tramways.

Sykes died at his home near Preston on 10th August 1907.

Board Members
Claud Thornton Cayley

Claud Thornton Cayley was appointed chairman of Dick, Kerr and Company, Limited in June 1890 when the company was first registered. In 1899 when the company went into voluntary liquidation and was restructured he became deputy chairman to John Kerr then in 1912, on the retirement of John Kerr, he took up the chairmanship again. In December 1918 he was one of founding directors of the English Electric Company Ltd.

He was born in Bickley, Kent on 9th June 1854 the son of John Cayley, and Clara Louisa Boulton, the daughter of Hon. Henry John Boulton; he died on 31st July 1928. He married, firstly, Esther Mary Bayley, daughter of William Bayley, on 1st June 1881, with whom he had six children; then in May 1927, he married Blanche Graham Ford, daughter of John Ford.

In 1911 he is living at 1 Evelyn Gardens, a mechanical engineer and Managing Director connected with the Admiralty, War Office.

Cayley was one of the proprietors and Managing Director of arms-manufacturing company, the Projectile Company (Limited) and from 1902 the Projectile Company (1902) Limited. In August 1892 he made a patent application 'for making tubular articles' (US 529597A).

In 1886 Cayley was a director of the Simonds Round-Forging Co. (Limited); in 1890 a director of the E. W. Bliss Co. (Limited), Brooklyn, U.S.A. who had acquired the American Patents of the Projectile Company (Limited).

Claud Thornton Cayley, Chairman of Dick, Kerr and Company Ltd. 1890 – 1899 & 1912 - 1919.

For recreation Cayley was a keen yachtsman and member of The Royal London Yacht Club. He served the club as Flag Officer for 24 years as Rear Commodore 1903 – 1904, Vice-Commodore from 1905 to 1920 then Commodore for the years 1921 - 1925.

Cayley had been educated in Canada where his father was Vice Commodore of the Royal Canadian Yacht Club and where he started sailing in his first boat 'May' a fifteen tonner. He joined the Royal London in 1896 and initially sailed the 45 ton 'Imatra', then in 1903 Cayley had built a 'unique' racing schooner, originally named 'Heartsease', later renamed 'Adela'. This was built by Fay & Co. to a W. C. Storey design and was 133 feet long, on deck, had a beam of 22 feet and displacement of 224 tons. In the Dover-Ostend race of 1908 –

'Adela, which is owned by Mr. Claud T. Cayley, the vice-commodore of the Royal London Yacht Club' made its racing debut for the season, competing in the A class. Adela was the winner in a time of 1h. 50m. 36s. winning a first prize of 1,000f.

Shortly before his death he sold 'Adela' to Sir H. Seymour King.

After his death a commemorative tablet was erected to his memory in Holy Trinity Church, Cowes. The tablet tells us that he – 'Crossed Over' 31st July 1928 and 'his ashes were committed to the sea nearby'. A competition trophy, the Cayley Memorial Challenge Cup, was presented by his family after his death; the 'Darings' still race in Cowes Week for the Cayley Memorial Cup.

The report of his death and Will published on 21st September 1928 –

Mr Claud Thornton Cayley, of 12, Hyde Park Gate, London, for some years Commodore of the Royal London Yacht Club, late chairman of Gooch's (Limited), and a director of the English Electric Co. (Ltd.) and of the Projectile and Engineering Co. (Limited), £140,826.

Lincoln Chandler

Lincoln Chandler was born in 1866 the son of William Chandler and Alice Britton, he married Elizabeth Wright.

With the formation of the United Electric Car Co. Ltd. in 1905 Lincoln Chandler, along with Ludford Charles Docker brother of Dudley Docker, was appointed to the board to represent the financial interests of the Metropolitan Amalgamated Railway Carriage and Wagon Company, Ltd. (MARCWC).

He was an 'official' of the MARCWC under the chairmanship of Dudley Docker and in 1907 was appointed a director on the board of BSA, then, in 1910, a Director of Daimler following the takeover by BSA. Another close associate of both Chandler and Docker was George Flett, managing director of Dick, Kerr and Co. Ltd. who was also a director of MARCWC and 'a key member of the Daimler board'.

George Flett

George Flett was born at Wick, Scotland in 1855 and became in 1885 the Dick, Kerr director concerned with the company's cable tramway contracts. In 1903 he became the Dick, Kerr director in charge of the contract to electrify the Liverpool-Southport Railway and subsequently became managing director of the parent Dick, Kerr Company. Mr. Flett had travelled widely in the service of Dick, Kerr and Co., visiting Japan and other Far East countries, Moscow, Egypt and Brazil.

Flett was a close associate of Frank Dudley Docker on the board of the Amalgamated Railway Carriage and Wagon Company Ltd. and was a 'key member' on the board of the Daimler Motor Co. of Coventry.

In 1904 George Flett was offered, and accepted, a seat on Mr Chamberlain's Tariff Commission. Reported by the Lancashire Daily Post on Monday 4th January 1904 –

> It was mainly through the influence of Mr. Flett, who was associated with Mr. John Kerr, M.P, that the electric works were installed at Preston. The company, which owns one of the largest electrical manufacturing concerns in the world, is at present engaged in constructing the new electrical tramways at Preston. Mr Flett, who is in the prime of life, has had a very large experience of the electrical industry, and has been actively and continuously identified with tramway developments in this country. He has travelled in India, Africa and America and is recognised as a leading authority in the particular business in which he is associated. Mr. Flett is a keen yachtsman. He was at Kiel last year, when both he and Mr. Kerr, M.P., were introduced to the German Emperor. 'Mr. Flett is not a politician'.

The Tariff commission was an unofficial body set up in 1903 in association with the Tariff reform league with Joseph Chamberlain as President and Sir Robert Herbert, G.C.B. as Chairman. The work of the commission would continue through the period of the Great War and ran until 1925.

George Flett had earned himself a considerable international reputation which is convincingly alluded to by a short article in the American publication, Street Railway Journal, of June 1906 –

> Mr. George Flett, managing director of Dick, Kerr & Company, Ltd., of London, left New York this week for London, after a stay of a few days. This will complete a trip around the world lasting six months. On 18th June Mr Flett was entertained to a farewell luncheon at New York, at which there were present Messrs. F. J. Spragne, J. G. White, H. H. Vreeland, H. McL. Harding, James H. McGraw, T. C. Martin, P. G. Gossler, J. M. Wakeman, H. G. Stott, M. G. Starrett, S. M. Curwen, W. S. Burnham, J. Goodell, A. N. Connette and W. S. Doran. He sailed from New York for England on 19th June.

> He had sailed from London for India in January and spent considerable time in the Far east where Dick, Kerr had built a number of tramway systems'.

The Times for Thursday 28th July 1910, reports –

'Fatal Motor Car Accident, Mr. George Flett killed'.

> A fatal motor-car accident took place at Birmingham yesterday. Three of the directors of the Metropolitan Amalgamated Railway Carriage and Wagon Company were travelling from the Saltley works, where a board meeting had been held, to New-street Station in a powerful Daimler car, and when it was travelling along Great Francis-street a child ran into the road in dangerous proximity to the car. The driver turned the car to get out of the way, but the movement was so sudden that the car ran into the kerb and was overturned. Mr George Flett, one of the occupants, of 39, Devonshire-place, London, W., who was seated in the front with the driver, was thrown into the roadway. The car fell upon him and he sustained serious injuries, from which he died shortly after his removal to the General Hospital. Mr. Flett's injuries were mainly internal.
>
> The other two directors, Messrs. Ludford Docker, the Metropolitan nominee to the UEC Board, and J. F. Cay, who were inside the car (a covered one), were badly shaken, but they managed to extricate themselves from the overturned car by breaking the glass. The driver was not hurt. The car struck the little boy, who it is feared, has sustained a fractured base of the skull.
>
> Mr. Flett, who was 55 years of age, held a prominent position in the electrical engineering and industrial world. He was the managing director of Messrs. Dick, Kerr and Co., of Preston, and he became a director of the Metropolitan Amalgamated Railway Carriage and Wagon Company soon after the amalgamation six or seven years ago. He was to have returned to London yesterday afternoon.

At the time of his death George Flett was, in addition to his position as managing director of Dick, Kerr, a director of the Metropolitan Amalgamated Railway Carriage & Wagon Company, Ltd.; the Newcastle and Gosforth Tramways; the Patent Shaft & Axletree Company, Ltd.; The Projectile Company; British Aluminium Company, Ltd.; Rio de Janeiro Tramway, Light & Power Company with recently opened offices at 46 Threadneedle Street, London; Mexican Light & Power Company Ltd.; Monterey Railway, Light & Power Company and the British Engineering Company of Egypt.

George Frederick Fry, J.P.

George Frederick Fry was on the board of directors of the Electric Railway & Tramway carriage works from its first meeting in May 1898. He was also a Chairman of four public companies and a director of seven others. He was director of a Land Company, Water Company and of three Tramway Companies, one in London, one in Belfast and one in Brisbane; he was also Chairman of the Metropole Hotel Company in Dover.

Fry was born on the 19th April 1829 in Bermondsey, London, the son of George Fry, a Builder and Ann (nee Chattin), he died on Saturday 19th January 1907 at his residence, 46 London Road Dover. In the early years of the 19th C his father, George, had carried out a good deal of important building work in Dover including the Market and Museum.

After leaving school Fry moved to London and became articled to a carpenter. At a later date he was articled for three years to John Wright, C.E., of Park Street, Westminster, the Engineer who constructed the South-Eastern Railway from Ashford to Dover. He then obtained an appointment as Clerk of Works under the Commissioners of Sewers for Westminster then, at the age of 25, he secured an appointment as Surveyor to six London Parishes, namely. St. Anne's, Soho, St. Paul's, Covent Garden, St. Mary's le-Strand, St. Clement Danes, the Liberty of the Rolls, and the Precincts of Savoy. During the long period he held this appointment he was associated professionally with many other important works in the Metropolis, representing such clients as the Princess Louise, the Duke of Argyle, the Duke of Bedford, and the Crown.

In 1879 he resigned all his appointments in London and returned to Dover. The following year he was elected to the Town Council as Conservative member for the Castle Ward, a position he held until his death. Fry became the Admiralty representative on the Harbour Board and was responsible for promoting a number of Harbour Bills which eventually made it possible to accommodate liners and other large ocean going ships. The widening of the streets of Dover, a project strongly supported by Fry, made it possible to introduce electric trams, he recognised the need for better public transport in Dover, and after a thorough analysis of the prevailing conditions he confidently proposed the adoption of an electric tramway.

As soon as the news of the death of George Frederick Fry was known, the flags on all public buildings were lowered to half mast, for, in addition to having been a member of the Town Council for more than 20 years, he was also a Magistrate for the Borough of Dover, and one of the Justices for the County of Kent. The Dover Express for Friday 8th March 1907 reported – The late Mr George Frederick Fry of Dover left estate valued at £32,902-16s-7d.

George Richardson

George Richardson, Iron Merchant, was appointed chairman of the Electric Railway & Tramway Carriage Works Ltd. from its incorporation in May 1898 then the United Electric Car Co. from its creation in 1905. He was chairman of the North Metropolitan Tramways Co., and the Blackpool & Fleetwood Tramroad Co., a director of the City of Carlisle Tramway and a trustee for the Mexican and Central American Railway Ltd. He was a director of many other tramway companies including those in the Provincial Tramways Group.

George Richardson was born in Glasgow in 1835, the son of William Richardson and Eliza Muirhead and died on 4th January 1918 aged 83; he had four siblings, two brothers and two sisters. In 1869 he marries Evelyn Collins, aged 25, the daughter of Charles and Sarah Collins of New York; they had four children, Frank Collins (1870 – 1917), Harold George (1876 – 1910), Hilda Evelyn (1878 – 1958) and Victor de Courcy (1884 – 1885). In the census returns George Richardson is an Iron Merchant with a home address in Paddington, in 1911 he is widowed and a retired Iron Merchant aged 75.

Walter Rutherford

Walter Rutherford joined Dick, Kerr and Co. in 1902, becoming managing director of Dick, Kerr Ltd. in 1910 following the death of George Flett, having previously been assistant managing director, then in 1918 he was one of the founding directors of the English Electric Company, resigning in 1922.

The Colliery Guardian dated 16th October 1936, reports –

Mr. Walter Rutherford, who has died in London at the age of 66, went to Canada in his youth and returned to this country at the age of 27 to become general manager of Dick, Kerr and Co. He afterwards became assistant managing director and then managing director.

Rutherford was one of a group of engineers whom George Flett drew together in the late eighteen nineties to gain for Britain a place in the field of electric traction, which was then almost wholly in American hands. He is credited with having played a major part in the successful achievement of that ambition and of making the name of Dick, Kerr known throughout the world for the excellence of their traction products.

At the conclusion of the Great War his energies were devoted to the formation of the English Electric Co., which combined the interests of Dick, Kerr & Co., the Phoenix Dynamo Manufacturing Co., Willans & Robinson, the Coventry Ordnance Works, and later, Siemens Brothers Dynamo Works. In 1922 he severed his connection with the English Electric Co., and started the Foundation Co. At the time of his death, he was managing director of the Metropolitan Construction Co., and a director of Balfour Beatty and Co. and the Calcutta Tramways, Ltd. He joined the Institution of Electrical Engineers as a Member in 1900, and served on the Council from 1907 to 1910 and again from 1912 to 1915. He was also one of the founders of the British Electrical and Allied Manufacturers Association.

Walter Rutherford died on 9th October 1936 at Westminster, estate £13,879-11-8.

Charles Edmond Webber

Major-General Charles Edmond Webber, CB, M. Inst. C.E., M.I.E.E., Royal Engineers, retired was a consulting electrical engineer and director of Dick, Kerr & Co. Ltd. from the registration of the company in May 1890.

He was born in Dublin on 5th September, 1838, the third son of Rev. Thomas Webber, of Leekfield, county Sligo; he was educated in private schools prior to his military education. Webber entered the Royal Military Academy, Woolwich in 1853 and was commissioned as Lieutenant in the Corps of Royal Engineers in 1855.

After receiving further instruction in the military art at Chatham, he returned to Ireland, where he was occupied on the construction of barracks at Lough Swilly. Webber was then sent to India where he served during the Indian Mutiny, 1857-1860. He was posted with the 21 company Royal Engineers who joined the 1st Brigade seeing action at sieges, Chandarie, Jhansie; battles Kota ka Serai, Betwa River, Koonch, Calpee, Gwalior, and many minor engagements. From April 1859 he served in the public works departments at Gwalior and Allahabad before returning to England in May 1860.

On his return to England he served in the Brighton district and it was here in 1861 that he married Alice Augusta Gertrude Hanbury-Tracy, fourth daughter of Thomas Hanbury-Tracy.

In 1866 he was instructor in topography, R.M.A.; with the Prussian Army; Paris exhibition in 1867; special commission preparatory transport Abyssinian expedition. In 1869, whilst in command of 22nd company Royal Engineers at Chatham, Webber and his men were lent to the Post Office to assist in constructing and organizing the telegraph service. In 1871 the 34th company was added to Webber's command and stationed at Inverness. More than 1,000 line miles and more than 3,200 wire miles were laid by them over and underground in 1871. In 1872 he was promoted to the rank of Major, then at the outbreak of the first Boer War in 1879 he accompanied Sir Garnet Wolseley to South Africa. Webber was promoted to the rank of Lieutenant Colonel in January 1882 and in August, appointed assistant-adjutant and Quartermaster General at the headquarters of the Egyptian Expedition; engaged at the Battle of Tel-el-Kebir; then in 1884-85 the Nile expeditions. He was mentioned in despatches six times; mutiny medal and clasp for Jhansie; African medal and clasp, C.B.; Egyptian expedition, 1882, medal and clasp and bronze star, 3rd class of the Medjidie for services in Egypt. In 1884 Webber was promoted to the rank of Colonel and made director of Army Telegraphs, until in 1885, he was declared supernumerary.

Whilst still in public service Webber served on the boards of the Bell Telephone Company, and the United Bell and Edison Telephone Company; he became a Director of the latter. Following retirement he became Consulting Engineer to the City of London Pioneer Electric Light Company, the Chelsea Electricity Supply Company, and the Anglo-American Brush Electric Light Corporation; he was a director of the latter two.

Following the death of Frederick Tollemache in1889, Webber was appointed to join his father-in-law as one of the Dysart Trustees.

Webber was founder, with the late Sir Francis Bolton, of the Society of Telegraph Engineers which in 1880 became the Institution of Electrical Engineers and of which Webber served as President in 1882. He published several works during the course of his career, on Military subjects (various), Telegraphy, Telephony, and on Electrical Engineering including, significantly, a contribution on Light Railways to the Encyclopaedia Britannica which was published several years after his death.

He lived at 17 Egerton Gardens, S.W. and was a member of the United Services Club.

Charles Edmund Webber died in Margate, Kent in 1904, aged 66, and is buried in the churchyard of St. Margaret of Antioch, Lee Green.

The Engineers
John. E. Anger

John E. Anger was born on 23rd June 1852 at Schenectady, New York to parents George Herman Anger, a German immigrant, and Hattie A. Smalley from Salem, Vermont.

His annual Certificate of Registration of American Citizen dated 14th February 1910, in Manchester, shows that he arrived in England on 3rd February 1899; and is residing at 186 Strand Road, Preston, for the purpose of business with the United Electric Car Co., Preston. His citizenship of the United States is established by 'Masonic Proof, Member of St. George's Lodge, No. 6, Schenectady, N. Y'. They had two sons Walter E. Anger, born 1886 at 1905 Market St. Wilmington, Del., and George H. Anger, born 1876 at Green Island, Albany County, N. Y. On 3rd September 1921 he makes an Emergency Passport Application to the American Consul in Liverpool giving his permanent address as Southport, England. The passport is required for the purpose of visiting the British Isles for 'residence arranging his affairs' and Belgium and France for 'commercial business'.

In the 1880 United States Federal Census, John, aged 27, Hattie, aged 26, and son George E., aged 3, are resident in Albany, New York; his occupation is that of Machinist. In the 1901 census John and Hattie are residing at 4, Mornington Road, Southport; his occupation is that of a General Foreman, Electric Tram Cars. In 1911, John aged 58 and Hattie aged 57, are at 186 Strand Road, Preston; his occupation is that of Works Manager, United Electric Car Co.

John Anger may have previously worked in Argentina, probably with the Anglo-Argentine Tramways. In July 1911 he tried to claim royalty on the 'semi-convertible' form of construction, but UEC would only pay such royalties direct to Anglo-Argentine. In 1912, John Anger devised an automatic brake shoe adjuster and assigned it to his two sons.

The Anger Manufacturing Co. Ltd. (Managing Director, George Herman Anger) was formed in March 1913, and UEC paid royalties to them. By 1916 J. E. Anger had set up his own company.

William Newby Colam, M. Inst. C.E., M. Inst. M.E.

William Newby Colam was born on 13th December 1853, at Hackney, to John Colam and Elizabeth (nee Newby), and died on 6th December 1930 at Hove, Sussex. Probate London on 7th February 1931 to Helen Maude Colam, Widow, and Stanley D'eyncourt Colam, Journalist.

William Newby Colam was probably the leading British authority on cable tramway systems with responsibility for most of the major projects of the 1880,s and 1890,s working with the Dick, Kerr Company and their chief engineer James More on a number of projects. He presented papers on the subject to the Society of Engineers in 1885, 'Cable Tramways' and to the Association of Municipal and County Engineers in 1890/91, 'Edinburgh Northern Cable Tramways', and again in 1897/98, 'Conversion of Edinburgh, Leith and Portobello Horse Tramways Systems into Cable Traction' both of which systems he designed and built. In 1891 Colam was President of the Society of Engineers.

His proposal for admission as a Member of the Institution of Mechanical Engineers, dated 10th September 1889, and aged 35, gives his occupation as that of Engineer and his business address as Cable Tramways Edinburgh.

A US patent application, filed on 7th April 1892, for a 'new and useful improvement in car couplings', (British patent number 2,620), has William Newby Colam, of London, England, assignor of one-half to Dick, Kerr & Company Limited and his residence as Billiter Buildings, in the city and county of London, England.

Sir Harold Nugent Colam, his son, was also a civil engineer, who was an important figure in the railway system of India.

Lieutenant-Colonel Patrick Walter D'alton

Patrick Walter D'alton was born in 1856 and was educated at Dublin and at King's College, London. His practical training was obtained at the Reading Iron Works, the North-Eastern Marine Engineering Works, and with Messrs. R. and W. Hawthorne. He subsequently went to sea and obtained his chief engineer's certificate. In 1886 he was appointed chief engineer to the Nordenfelt Submarine Boat Co.

A few years later he turned to electrical engineering and in 1889 became resident engineer at the Deptford power station of the London Electric Supply Corporation. In 1892 he succeeded Mr. Ferranti as engineer-in-chief of that company. On the outbreak of the South African War he took up military service and rose to the rank of Major in the 6th Battalion of the Royal Irish Rifles. At the end of 1900 he became chief engineer to the Central London Railway, and then in 1903 he joined the staff of Dick, Kerr & Co. Ltd., as chief mechanical engineer, remaining with the Company until November 1914, when he set up in private practice as a consulting and inspecting engineer. In 1916 he became chief engineer to the Valuation Committee under the Ministry of Munitions with the rank of Lieutenant-Colonel, holding that office until the Committee was dissolved in 1921. He then resumed private practice, and died on 10th April 1927, aged 71.

Professor Samuel Byrod Fortenbaugh,

M.M.E., M. Am. Inst. E.E.

Samuel Byrod Fortenbaugh was born in 1865 in the village of Halifax, Pennsylvania, USA. He prepared for college in the High School at Ithaca, N.Y. and entered Cornell University with the class of 1890. After graduating from the electrical engineering course he was employed by the Brush Electric Co. of Cleveland, Ohio; two years later he resigned his position with that firm to become assistant chief engineer of the Short Electric Railway Co. of the same city. He subsequently returned to Cornell for graduate work, and received the degree of Master of Mechanical Engineering.

In 1894 he became Assistant Professor of Electrical Engineering at the University of Wisconsin, working with Dugald Caleb Jackson, Professor of Electrical Engineering, who had also trained at Cornel. In 1895 Professor D. C. Jackson directed Assistant Professor Samuel B. Fortenbaugh to develop a division of applied electrochemistry within the Department of Electrical Engineering, marking the beginning of chemical engineering at UW-Madison. In 1895, and in collaboration with D. C. Jackson, Fortenbaugh published a paper 'Some Observations on a Direct-Connected 300 kW Monocyclic Alternator'.

Having apparently distinguished himself in the field of education his future career was, nevertheless to lie outside the Engineering College.

News from the Alumni, University of Wisconsin, for 1901 provides the following 'Personal Notes' –

> Prof. Samuel B. Fortenbaugh was married January 10th to Miss Florence Cowden at the Pine Street Presbyterian Church, Harrisburg, Pa. The couple will be at home after February 15th at Knowle, Edgerton Road, Ashton-Ribble, Preston, England. Professor Fortenbaugh left the University two years ago to enter the service of the Walker Electrical Company of Cleveland and will now represent a company in England that is operating under American patents.

When he came to England from America, S. B. Fortenbaugh had the reputation of being one of the 'most brilliant electrical engineers in the World' having gained experience on the Chicago transit systems. *It would appear that Fortenbaugh is working very closely with Sydney Short having previously worked for him in Cleveland.*

When Sydney Short, accepted the appointment of technical director with the newly formed English Electric Manufacturing Company in Preston he recruited S. B. Fortenbaugh, his colleague at the Walker Company, to be his assistant.

Samuel B. Fortenbaugh was closely involved with the work carried out by Dick, Kerr for London County Council including the electrification on the Thames Embankment. His remuneration is discussed in the early minute books but his period of employment with the company was a relatively short one for in 1904, from the listing of shareholders, he is seen to be living in London and employed as an Electrical Engineer with the Underground Electric Railway of London (UERL).

In 1906 he publishes a report, in the Street Railway Journal, 'The Electrification of the London Underground Electric Railways Company's System' as Electrical Engineer for that Company. The Chief Engineer and General Manager of the UERL was another American, James Russell Chapman and from 1902 the Chairman also an American, Charles Yerkes, seems to have had a rather chequered career financing the tramway system in Chicago.

In 1906 Fortenbaugh returns to the United States to join the traction department of the General Electric Company at Schenectady.

Professor Harris Joseph Ryan

The association between Professor Ryan and the English Electric Manufacturing Company appears tenuous and is only known about from a single quote attributed to Mr. R. G. Devey who joined the company in 1900 when the new works were under construction. Mr. Devey was interviewed for the English Electric Journal, English Electric and its People, in 1966. There is no extant record of Professor Ryan having worked for the Company but he would have been known to Professor Fortenbaugh, for three reasons: firstly Ryan was married to Catherine, the sister of S. B. Fortenbaugh; secondly they both had roots in Halifax, Pennsylvania and thirdly they both studied at Cornell University.

Harris Joseph Ryan was born on 8th January 1866 in Matamoras, Pennsylvania to Charles W. and Louisa M. Ryan and was married on 12th September 1888 to Katherine E. Fortenbaugh in Halifax Pa.

Ryan entered Cornell University in 1883 to study electrical engineering in the Department of Physics under Professor William A. Anthony. After graduating, in 1887 he was offered an instructorship in physics; but Ryan decided to go into partnership with two of his contemporaries, J. G. White and Dugald C. Jackson (*also a connection to Fortenbaugh*) to form the Western Electric Company in Lincoln, Nebraska.

After only one year Ryan returned to Cornell to resume his academic career as Instructor in Physics and in 1891 Jackson was called to the University of Wisconsin to establish a department of electrical engineering.

In 1889 Ryan was appointed Assistant Professor of Electrical Engineering in Sibley College. In the autumn of 1905 he accepted an offer from Stanford University, California to head the Department of Electrical Engineering; he remained in that position until his retirement as Emeritus Professor in 1931.

Ryan died on 3rd July 1934 at his home in Palo Alto.

Professor Ryan, possibly in accordance with the business environment of the college, was identified as 'consulting engineer' and it may have been in this capacity that he became involved, perhaps for only a short period, with the English Electric Manufacturing Co., after all, it would also have been a good excuse for visiting his brother-in-law in Preston.

During his career Professor Ryan achieved considerable eminence in his chosen fields of research, including: transformers, the 'Conductivity of Atmosphere at High Voltages', high voltage transmission phenomena and power losses in high voltage transmission lines due to the corona effect. He was awarded the Edison Medal of the AIEE in 1925 for his contributions to the science and the art of high tension transmissions of power.

Professor Sydney Howe Short

Sydney Howe Short was born in Galena, Ohio in 1858 and grew up in Columbus, where his father was a manufacturer. He developed an early interest in electricity becoming a telegraph operator at the age of 14 then studied physics at Ohio State University. Graduating in 1880, he accepted a professional position at the University of Denver, where he taught physics and chemistry and pursued research in electrical applications, becoming Vice President of the University of Denver and Professor of Physics and Chemistry.

He built a 100-yards circular tramway, in the grounds of the University, and worked it on a 'series system'; after demonstrating his invention of the electric trolley car, local capitalists organised the Denver Tramway Company to build an electric streetcar in that city. A full size line was worked in the streets of Denver in 1886-7, but reverted to horse working, while a trial line laid at Northfleet, Kent, fared little better.

Obtaining a contract for a trolley system in St. Louis, he secured financial backing from an Ohio investor who arranged to have the necessary dynamos custom-made at the Brush Electric Company. Short then moved to Cleveland to supervise the work and experiment in the company's shops. Brush not only encouraged Short's efforts but helped to finance the resulting Short Electric Railway Company, which operated out of a Brush building. Short sold the company to General Electric in 1892 and then re-entered the business almost immediately by joining forces with the Walker Manufacturing Company, a cable and machinery builder, to develop an electric traction system. Walker Motors and Controllers were fitted to tramcars built by Milnes and then Brill and supplied to Dover Corporation in 1918.

Short served as their chief engineer and designer then vice president until Westinghouse bought control in 1898.

That same year Short received an invitation from the Equipment Syndicate Ltd. of Manchester, with whom he agreed access to his valuable patents, to join with them in a venture to establish a new electrical manufacturing works in Preston for the manufacture of equipment previously imported from the United States and Germany. When, in 1899, Dick, Kerr & Co. took over control of the new works on Strand Road, Preston, to found the English Electric Manufacturing Co. Ltd., he was offered the position of Technical Director. Sydney H. Short continued to take out Letters Patent on numerous inventions; examples include – Automatic Cut-Out for Dynamo-Electric Machines, September 1st 1891 and Motor Mechanism for Electric Cars, September 22nd 1891.

His passport application dated 25th November 1898, for he and his wife, gives his description as – age 40 years, 5ft 11½ inches tall, high forehead, hazel eyes, Roman nose, firm mouth, strong chin, dark brown hair and medium fair complexion. He gives his occupation as Electrical Engineer and his intention to return to the United States in about six months to resume his citizenship therein. He must have thought his visit to England would be temporary, when in fact he would be taking up permanent residence in England and his return visits to the United States would be of short duration.

Sydney Howe Short died in October 1902, aged 44, at home in London, Middlesex, estate £18,821-6-1. He left a widow and four children and bequeathed the patent rights on about 500 inventions and an estate worth about 2, 5 million dollars.

An obituary published in the Street Railway Journal dated 25th October 1902 gives an interesting American slant to his life and work, quote –

Death of Prof. Sidney H. Short –

A cablegram from London, dated Oct. 22, announces the death in that city from appendicitis, of Professor Sidney H. Short, technical director of Dick, Kerr & Co.

Professor Short was one of the pioneers of the electric railway business and until his departure from this country had been prominently and continuously identified with the development of the electric railways of America. He was born in Columbus, Ohio in 1857 and was graduated from the Ohio State University in 1880. While in College he invented and patented a long-distance telephone transmitter and an improved arc lamp. After graduation he was appointed professor of physics and chemistry in Denver University, and while in that city built, in 1885, his first electric railway, which was a conduit line. In 1887 he returned to Columbus and built a 2½ mile electric railway in that city. He built another line in St. Louis in 1888, and in 1889 removed to Cleveland where he organised the Short Electric Railway Company, which soon took a prominent place as a manufacturer of electric railway apparatus.

After the absorption of this company by the General Electric Company Mr. Short took a short rest, but was soon invited to engage in his chosen field by the Walker Manufacturing Company which up to that time had been a large manufacturer of cable machinery. The managers of this company decided to engage in the manufacture of electrical apparatus, and selected Professor Short as its vice president and electrical engineer. Walker motors and generators were soon placed upon the market and achieved a high reputation for excellence of design and construction. While connected with this company Professor Short paid special attention to the problems of heavy electric railroading and developed an 80-hp motor for electric railway work, which was adopted on the Brooklyn Elevated Railway, and a 150-hp motor, which was put in service on the Metropolitan West Side Elevated Railway of Chicago. While connected with the Walker Company Mr. Short also developed a pneumatic system of multiple unit control.

In 1898 the control of the Walker Company was secured by one of the other large manufacturing companies, and early in 1899 Professor Short sailed for London, where he accepted the position of technical director of the English Electric Manufacturing Company, a corporation which had recently been organised by British capitalists associated with Dick, Kerr & Co., to build electrical apparatus of all kinds, but particularly for street railway service.

Under the direction of Professor Short very large works were erected by the English Electric Manufacturing Company at Preston, England, and the work of manufacture was immediately commenced. The success of the enterprise was immediate from the start, and orders were received not only from many of the large tramway companies in Great Britain but also from British colonies in the East and from a number of continental countries. Professor Short's record in England as an electrical inventor and manufacturer was equally as successful as that in America, and at the request of a number of scientific bodies he presented papers before them on various technical subjects connected with the design of electrical apparatus. In spite of his residence abroad he never gave up his American citizenship and with his wife made several trips to this country, the last one being on the occasion of the convention of the American Street Railway Association in New York in 1901, where he renewed his acquaintance with many of his former friends.

As an inventor Professor Short's record stands extremely high, and has been marked by progressiveness, as he has kept in touch with the latest developments in electric railroading, and has made improvements and inventions in advance of the art. Personally he was an extremely agreeable companion and had a very wide circle of friends. Professor Short leaves a wife, three sons and one daughter. His eldest son is now taking the engineering course at Cornell University.

On 8th November 1902 the Street Railway Journal published a further item on the death of Sidney Short –

Electrical circles in London and in fact throughout the world, have been deeply moved to learn of the untimely demise of Professor Sidney H. Short, who has for the past three or four years been acting as technical director of Messrs. Dick, Kerr & Co., of London. Mr Short had established for himself an enviable reputation in England in connection with the almost unprecedented success and phenomenal growth of Dick, Kerr & Co.'s electrical business and had surrounded himself with a large and influential circle of friends. Mrs Short, who has always taken an active interest in Mr. Short's success, had also made a social position for herself in London, and has our most sincere sympathy in her sad bereavement.

Arnaldo Paolo Zani

Zani was one of the Technical specialists recruited from the United States for employment by the English Electric Manufacturing Co. Ltd., and working alongside Short, Zweigbergk and Fortenbaugh. He was born in Italy in 1872 and sailed to the United States from the port of Genoa arriving in New York on 16th May 1895, aged 23. In 1911 he is living at 42 Haverstock Hill, London; Arnaldo P. Zani, age 37, of Italian nationality, born in Italy and single, occupation Civil & Electrical Engineer working in the Electrical Industry.

Zani had a brother, a Dr. Zani, who designed the 'Filovia' system of overhead current collector used by the United Electric Car Co. for a trolleybus built in about 1911, and electrically equipped by Dick, Kerr & Co. Ltd. The 'Filovia' system was made by Societa per La Trazione Electrica of Milan and licensed to a Manchester firm, UEC operating out of Trafford Park, they went bankrupt in 1911.

Thorsten Von Zweigbergk

Thorsten Von Zweigbergk a subject of the King of Sweden and Norway was born at Sandby, near Lund, Sweden on 19th August 1871, the son of Axtel Von Zweigbergk and Malin Hylen; he married Katherine McNally at Cuyahoga County, Ohio on 28th June 1899.

In 1898 Thorsten Von Zweigbergk, is residing in the county of Cuyahoga and State of Ohio and working for the Walker Electric Co. of Cleveland, Ohio. He would therefore have been a colleague of Professor Short, accompanying Sydney Short to England and becoming the 'chief trouble-shooter' at the Preston Works. After Professor Short's untimely death in 1902, Zweigbergk becomes Technical Director responsible for all Dick, Kerr traction work, culminating in the 3,000 volt dc electrification of the railway from Bury to Holcombe Brook. He retired in the early 1920s and became a consulting engineer, and in 1929 shared in the design and development of the Liverpool monomotor bogies. He died in Stockholm on 11th February 1930.

From his United States Passport application of 15th May 1899 he emigrated to the United States on the Ethiopia from Glasgow in August 1892 and had resided there uninterruptedly for seven years from 1892 to 1899, at Cleveland, Ohio. He had been naturalised as a citizen of the United States before the Common Pleas Court of Cuyahoga County, Cleveland, Ohio on 15th May 1899; his address is given as Lancaster, England and his occupation as Electrical Engineer.

His travels between England and the United States give some indication of the difficulties that the authorities had with his name; on 5th April 1906 from Liverpool to New York on the Carmania he is Zuegberg and on 29th February 1908 from Liverpool to New York on the Mauretania he is Zwergbergk.

Thorsten Von Zweigbergk continues to make US patent applications during his time with Dick, Kerr & Co. including one for a Motor Control System on 16th January 1911 for which he states that he is 'of Southport, England, Assignor to the General Electric Company, a Corporation of New York'. Thorsten Von Zweigbergk had at least 26 patents applications to his name or jointly with his brother Gillis.

For the 1901 census he is resident at the Crown Hotel, Miller Arcade, Preston, age 30, and born in Sweden; with him his wife Catherine, age 27, born in the U.S.A. In 1911 he is resident at 20 Allerton Road, Southport, age 40, an Electrical Engineer, born in Sweden and a citizen of the U.S.A. With him are his wife Katherine, age 35, born in the U.S.A and two servants. The census return reveals that they have had two children and that they have both died.

Thorsten Von Zweigbergk had been joined in England by his brother Gillis Von Zweigbergk, also a talented electrical engineer, and they worked together for the English Electric Manufacturing Co. Gillis had a son Tony Zweigbergk, born in Preston on 22nd April 1916. From Tony's War: the Life and Times of a WW2 Typhoon pilot by Britta Von Zweigbergk, Thorsten and Gillis had spent some years in America before finally settling in England. Life was not without its difficulties – not least having a Germanic sounding name. This was not helpful in the early twentieth century in connection with WW1 as in later years. The Von was dropped from the name and not reinstated until the late 1950s. Tony Zweigbergk joined the RAFVR and when war broke out in 1939 he joined the RAF where he had a distinguished war career as a fighter (Typhoon) pilot rising to the rank of Squadron Leader. He played his part after the war in helping countries such as India and Burma develop their own airlines and worked with Silver City Airways at Lympne and Ferryfield.

The Managers
James Conner

Was born in Glasgow in 1858 and was educated at the Ayr Academy and the Andersonian College. He became an apprentice at the St. Rollox locomotive works of the Caledonian Railway and then joined the Hyde Park Locomotive Works at Springburn. In 1884 he became Chief Draughtsman at the Britannia Works of Dick, Kerr & Co. at Kilmarnock. In 1891 he was living at 10 Hill Street, Kilmarnock with his wife Susan and three children, occupation Draughtsman.

In 1901, age 45, he is at 4 Clark Street, Kilmarnock with his wife Susan and four children, occupation Engineer (Works Manager). By April 1902 he has transferred to Preston as Works Manager of the recently established English Electric Manufacturing Co., on Strand Road, a subsidiary of Dick, Kerr and Co. James Conner was in the forefront of the application of steam tramways, in the use of large gas engines in industry, and in the production of electric tramways and railway material in this country. His management of the Preston works of Dick, Kerr & Co., coincided with the rise of that firm in the electric tramway and railway world, and he bore the responsibility of converting that works into an armament factory in the first year of the Great War.

In 1911 he is living at Egerton House, Preston with wife Susan and four children. He was elected a member of the Institution of Electrical Engineers in 1912 and retired in 1920, becoming a consulting engineer and a director of the Kilmarnock Engineering Co. He died on 22nd May 1937 at the age of 79.

Lewis Edwin Pickin

Lewis Edwin Pickin was the Company Secretary for the Electric Railway & Tramway Carriage Works Ltd., and the United Electric Car Co. Ltd. His minute books from 1898 to 1916 give the impression of a man who has a total grasp of the business; his correspondence, both internal and external, and particularly during periods of change within the company, clearly indicate that he has the complete confidence of the chairman and the board.

He was born in 1872 at Clifton, Gloucestershire and died on 21st March 1959, aged 86, in Marylebone; probate 25th May 1959 London, Lewis Edwin Pickin of 24, Charlbert Court, St. Johns Wood, £3,368-2-4; he was married in 1902 to Nellie Bell Mulock, registered on the Fylde.

In 1881, aged 8, he is at Westbury upon Trym, Bristol with parents George G. Pickin, a bank manager, and Sarah Ann and five siblings. In 1911, aged 38, he is residing at 19, West Cliff, Preston, his occupation is that of a chartered accountant and secretary United Electric Car Co. Preston, a manufacturer of electric tramcars and railway carriages; with him are his wife Nellie Bell, two daughters and a servant.

E. A. Stanley

E. A. Stanley was recruited from the United States to become Car Works Manager at the Electric Railway & Tramway Carriage Works. He had been with General Electric but at the time he was 'head hunted' he was working for Jackson and Sharp. Among the ideas he brought with him was a design for the plain arch ceiling, which the Preston Works first used in 1899 on a batch of cars for Liverpool Corporation, and which replaced the complicated clerestory ceiling then in common use.

Solicitor and Trustee
Thomas Strover Turnbull

Thomas Strover Turnbull was a partner in the firm of Sutton, Elliott, Turnbull and Mayne, Solicitors of Spring Gardens, Manchester, Solicitors for the Electric Railway and Tramway Carriage Works Ltd., with a financial interest in the Company. Along with Richard Henry Prestwich he was a debenture trustee for the United Electric Car Company Ltd.

He was born on 17th March 1853 in Sunderland, County Durham, the son of Thomas Scott Turnbull a Draper of High Barnes, Sunderland. From humble beginnings, Thomas Scott Turnbull had built up his Sunderland-based business 'Albion House' into one of the largest drapery houses in Northern England; his premises included sleeping and dining accommodation for 160 assistants, plus a library of nearly 2,500 volumes. Politically he was a Liberal and in 1866 was elected to Sunderland Town Council, but retired three years later. In partnership with Samuel Storey, who was also his business partner and future MP for Sunderland, he co-founded the Sunderland Echo newspaper.

Thomas had a Private School education then went on to Trinity Hall, Cambridge being admitted on 19th September 1871, matriculated Easter 1872 and gained his BA in 1875, he was a rowing blue in the University boat race winning team of 1873. He was admitted a Solicitor in December 1877 and practiced in Manchester from 1878 with the firm of Sutton, Elliott, Turnbull and Mayne. Thomas married Fanny Pearce Lloyd, daughter of Thomas Lloyd, on 5th September 1878 at Baguley, St. John; he died on 5th October 1917.

ACKNOWLEDGEMENTS:

It would be impossible to carry out any research into the history of a company from this period without assistance from many people and organisations. The nature of the business also necessitated reference to published records, including the work of tramway and early railway enthusiasts who, over the years, painstakingly recorded, in great detail, the history of many early tramways and electric railways including the manufacturers, listing their products and customers.

Libraries and Record Offices, and the records they hold, were an invaluable source of information, particularly the Lancashire Archive where all the extant records of the Strand Road Works are preserved, including early correspondence, minute books and photographs from the Dick, Kerr archives.

The Harris Museum and Art Gallery in Preston and Emma Heslewood very kindly provided copies of the original 1899/1900 photographs of the English Electric Manufacturing Company works, on West Strand, under construction (about 60 images).

I am also indebted to the following:-

- Preston City Council and Harry Seddon.
- Lancashire Libraries, the Harris Library, Preston and the friendly assistance of all their staff.
- Lancashire County Council Cultural Services for photographs from the period of the Great War.
- Blackburn Central Library and the friendly assistance of all their staff.
- Gail J. Newsham for photographs and information on the Dick, Kerr Ladies Football Team.
- Birkenhead Library and the friendly assistance of all their staff.
- The Tramway and Light Railway Society and John Prentice for permission to quote from the works of the late J. H. Price and other publications and for allowing the use of photographs from the archive.
- Dave Jones and the London County Council Tramways Trust for providing and granting permission to reproduce photographs from their collection.
- The Industrial Railway Society and Russell Wear for permission to quote from 'The Locomotive Builders of Kilmarnock', Industrial Railway Record No 69 January 1977.

- David & Charles, publisher, for permission to quote from 'Light Railways of the First World War' by W. J. K. Davies and 'The Lancashire and Yorkshire Railways Vol. 2' by John Marshall.
- The Royal London Yacht Club and Rodney Barton for kindly providing me with the Portrait of Claud Thornton Cayley.
- Barney Smith Content Director, Preston Digital Archive for permisssion to download images from the archive.
- The Lancashire and Yorkshire Railway Society.
- James Dowswell for providing me with personal documents concerning the career of his Great Uncle Robert Livingstone.
- My friends John and Val. Bishop for research into the W. B. Dick connection with the Spa Hotel and Golf Club at Tunbridge Wells.

I would also like to express my appreciation of the help and support I received from my colleagues within the Heritage Department, especially the late James Longworth for encouraging me to continue with the project; also Tony Wilson for obtaining copies of reports held by the National Archive, Gerald Wilson for introducing me to James Dowswell, Dave Banks and Joe McGrath for finding the occasional gem of information and Allan Seabridge and Dave Ward for proof reading the early draft manuscript. I am particularly grateful for the efforts of Ian Lawrenson, who manages the Heritage Department in addition to carrying out his other duties within BAE Systems, for supporting and facilitating the preparation of the work, including liaison with the Printers and the relevant departments within BAE Systems.

Finally, my thanks to Gavin Hawthorne and Tim Harwood of Reid Creative in Preston, for patiently accepting my changes and corrections during the final stages of preparation for publishing and for their superb graphic design input.

While acknowledging all the assistance, information and material I have received, the final content of this publication is based entirely on my own interpretation of that material.

John Shorrock
January 2016

ILLUSTRATIONS:

In addition to the acknowledged sources, and in the absence of original images, illustrations have also been reproduced or downloaded from the following:

Publications produced by or for English Electric, including English Electric and its People (company newsletter), Electricity in Transport (1951), Early Histories of some companies of the English Electric Group(c1950) and the Dick, Kerr Album, a souvenir of Preston Guild (1972).

Other sources include - The Street Railway journal, the Tramway Review, the Engineer journal, Northern Daily Telegraph, Lancashire Daily Post, the Fire Appliance Company brochure, Jacqueline Banerjee and the National Railway Museum, New Delhi, the Victorian Web (Locomotive MTR No. 2) and early postcards.

Every effort has been made to trace copyright holders and to obtain permission for the use of sourced material.

The author apologises for any errors or omissions and would be grateful to be notified of any corrections that should be incorporated.

BIBLIOGRAPHY:

- The Dick, Kerr Story by J. H. Price 1993.
- The Story of G. F. Milnes by J. H. Price 1964.
- The British Electric Car Company Limited by J. H. Price 1978.
- Early Histories of Some Companies of the ENGLISH ELECTRIC GROUP c1950.
- The Story of the Preston Works by Ray Abbott.
- Electricity in Transport by H. H. Andrews, English Electric 1951.
- British Steam Locomotive Builders by James W. Lowe 1975.
- The Times Newspaper.
- The Engineer journal (website).
- Flight International Magazine.
- The Tramway and Light Railway Society.
- Light Railways of the First World War by W. J. K. Davies.
- The British Internal Combustion Locomotive 1894 – 1940 by Brian Webb.
- LCC Electric Tramways by Robert J Harley.
- London Trams – A View from the past by Paul Collins.
- Farnworth – The Archive Photographs series - compiled by Ken Beevers.
- The Lancashire & Yorkshire Railway Volumes 1 & 2 by John Marshall.
- Tony's War: the Life and Times of a WW2 Typhoon pilot by Britta Von Zweigbergk.
- Personal Capitalism and Corporate Governance: British Manufacturing in the first half of the Twentieth Century by Matthews, Maltby, Lewis & Rogers – Ashgate Publishing.
- Networks of Power: Electrification in Western Society 1880 – 1930 by Thomas P. Hughes.
- Alford & Sutton Tramway by George Dow 1984.
- Tramways of the City of Carlisle by George S. Hearse 1962.
- The University of Wisconsin 1848 – 1925, Merle Curti & Vernon Carstensen.
- The Locomotive Builders of Kilmarnock, Russell Wear 1977.
- The Cable System of Tramway Traction, first published 1896, republished 1994.
- Cable Tram Days by F. K. Pearson, T. Eng. (C.E.I.), Douglas Cable Car Group, 1977.
- Volk's Railway Brighton by Alan A. Jackson, 1976.

- The British Horsetram Era, E. R. Oakley, Tramway & Light Railway Society, 1978.
- Triplane to Typhoon by James H. Longworth.
- The Story of Dover Corporation Tramways, 1897 – 1936 by J. V. Horn, 1955.
- Street Railway Journal.
- The Tramway and Railway World.
- English Electric and its People (Company newsletter).
- Manchester Grammar School Magazine, October 1917.
- The Wirral Archive.
- R. J. Mitchell, edited by Gordon Mitchell.
- The Dick, Kerr Album, a souvenir of Preston Guild 1972.
- The Rothesay Tramways Company 1879 – 1949 by In L. Cormack.
- The Tramways of Kent – Vols. 1 & 2 by Invicta.
- Edinburgh Transport by D. L. G. Hunter, M.I.Mech.E, M.Inst.T.
- The Metropolitan Electric Tramways – Vol. 1 – Origins to 1920 by C. S. Smeeton.
- Tramways by D. Kinnear Clark, C.E., M.Inst.C.E; M.Inst.Mech.E, 1894, reprint No.5 1992 version.
- Business History 1985 Vol. 27 Issue 1, Liverpool University Press.
- The History of the J. G. Brill Company by Debra Brill.
- Locomotives of the Namaqualand Railway and Copper Mines by Peter Bagshawe.
- A Century of Technology edited by Percy Dunsheath C.B.E.
- The Tramway Review Vol. 4, Issue No. 25, 1958.
- In a League of Their Own! The Dick, Kerr Ladies 1917-1965 by Gail J. Newsham.

APPENDIX 1:

In Memory

Of those Men from Dick, Kerr's Preston works who were killed in the Great War, 1914 – 1919.

Adams, R – King's Liverpool.

Allison, James – South Lancs.

Ashton, Harry – L.N.L.

Ashworth, R – L.N.L.

Aspinall, R – L.N.L.

Atkinson, F – L.N.L.

Atkinson, H. E. – West Ridings.

Atkinson, J – L.N.L.

Atkinson, Stanley – L.N.L.

Barker, J – L.N.L.

Barnes, E – R.E.

Barnes, T - L.N.L.

Bateson, W - L.N.L.

Bennett, James – S. W. Border.

Bennett, Thos. - L.N.L.

Bilsborough, R - L.N.L.

Bird, J. Sergt. – Gordon Highl'rs *(Died of wounds).*

Blue, J – H.A.C.

Bond, R – Lancashire Fusiliers.

Bond, R. O – R.E.

Boote, J. P - L.N.L.

Boothby, J – R.E.

Bowes, J - L.N.L.

Boyd, S – R.F.A.

Bramwell, W – Royal Scots Fusiliers.

Bruford, R – Lancashire Fusiliers.

Burscough, Geo. F – R.F.A.

Butler, H. Lieut. – R.A.F.

Butterworth, J - L.N.L.

Chapman, W – Welsh Regiment.

Charnley, J - L.N.L.

Charnock, J – King's Liverpools.

Charnock, J - L.N.L.

Chevelier, V – M.G.C.

Clarkson, H, Sergt. - L.N.L.

Clarkson, W – R.F.A.

Cooke, F – M.G.C.

Cookson, J – West Yorks.

Cookson, J – Lancs. Fusiliers.

Cookson, Jas. – R. Inniskillen F.

Cookson, M – K.R.R.

Corless, H - L.N.L.

Corless, J – Scots Guards.

Coupe, R - L.N.L.

Coulthurst, I – Scots Guards.

Cowley, J – R.E.

Crozier, H – M.G.C.

Dawson, Herbert – K.O.R.L. *(Died of wounds).*

Denver, J - L.N.L.

Devey, F. Sergt. - L.N.L. *(Died of wounds).*

Dixon, W – L.N.L.

Duffy, John – L.N.L.

Durham, W – L.N.L. *(Died of wounds).*

Fitzsimmons, Thos. Sergt. – Manchesters.

Fletcher, J – L.N.L.

Flynn, W – L.N.L.

Foster, J – East Lancashires.

Fraser, A – L.N.L.

Garlick, L – S. Wales Borderers.

Gillibrand, J – L.N.L.

Goring, W – L.N.L.

Gregg, C. Lieut. – K.O.R.L.

Gynes, C. F – L.N.L.

Hacking, W – South Lancashires.

Hall, E – Scottish Rifles.

Hall, J – L.N.L. *(Died of wounds).*

Hall, Wm. – L.N.L.

Hart, G – Argyle & Sutherland Highlanders.

Hartley, George – King's Own Lancasters.

Helm, Fred. – Sergt. – East Lancashires.

Henery, W – L.N.L.

Higham, R. M. – L.N.L. *(Died of wounds).*

Hobson, J – L.N.L.

Hockey, F – R.F.A. *(Died of wounds).*

Hodson, C – R.F.A. *(Died of wounds).*

Hodson, J – Gordon Highlanders.

Holdway, C – L.N.L.

Holt, A – R.F.A.

Howe, G. H. Lieut. – R.G.A.

Ibbison, E – L.N.L.

Ibison, J – R.F.A.

Jackson, J – L.N.L. S

Jackson, R – Lancs. Fusiliers *(Died of wounds).*

Jackson, W – L.N.L.

Jemson, T, Corporal – L.N.L.

Johnson, Percy – L.N.L.

Johnson, W – "Lobelia 2nd" *(Minesweeper).*

Kay, H – Manchesters.

Keen, W – L.N.L.

Kettlewell, A – R.A.F. *(Died of wounds)..*

Knowles, J – L.N.L.

Lawrenson, F. Sergt. – L.N.L. *(Died of wounds).*

Lawrenson, G – King's L'pools.

Leeming, Jas. – King's L'pools.

Leonard, J – Liverpool Irish.

Lindsay, Arthur – Scots Guards.

Maher, Thos. – K. L'pools (Died).

Maynard, J – L.N.L. Woods, B – L.N.L.

McCartney, A – East Lancs. *(Drowned at sea).*

McGovern, J – L.N.L.

McIntosh, C. Corporal – R.F.A. *(Died of wounds).*

McLaughlin, James – L.N.L.

Moizer, Edward – R.F.A.

Ogden, J – Royal Inniskillen Fusiliers.

Oldway, Chas. – L.N.L.

Parker, R – M.G.C.

Parker, T. R – L.N.L.

Parker, W – L.N.L. *(Died of wounds).*

Partington, W – R.F.A.

Pettitt, F – R.D.C. (Died).

Phillipson, S – Royal Fusiliers *(Died).*

Pickering, L. Capt. – R.E.

Poole, H – L.N.L.

Preston, T – L.N.L.

Prince, J – King's Liverpools.

Rathbone, J – L.N.L. *(Died of wounds).*

Raynor, John – L.N.L. *(Died of wounds).*

Robinson, A – R.E. *(Died of wounds).*

Robinson, T – L.N.L.

Rogerson, B – L.N.L.

Rogerson, A – L.N.L.

Sharples, J – A.S.C. *(Died).*

Shaw, Joe – King's Liverpools.

Spiby, T – Household Batt.

Smith, A – L.N.L (Died).

Stephen, W. H – L.N.L.

Stephens, W. Sergt. – L.N.L.

Street, R – L.N.L.

Syddall, H – L.N.L.

Taylor, R – Lancashire Fusiliers.

Thompson, J – R.E.

Thompson, Fred. – Welsh Regt.

Thorpe, A. S. – L.N.L.

Tuson, R – Royal Naval Division (Died).

Utting, Wm. – L.N.L. (Died).

Walker, Edgar – Cheshires (Died).

Walmsley, J – L.N.L.

Woods, R – L.N.L.

Whittle, G – Royal Fusiliers.

Worden, J – Cheshires.

Wright, J – L.N.L.

Wright, William – Lancs Fus.

Williams, H – A.S.C. (Died).

Woods, B – L.N.L.

Whittaker, S – R.E.

Webb, Tom – Cheshires.

Whalley, J – L.N.L.

Whatnough, T – L.N.L.

Whelan, F – Coldstream Guards.

Abbreviations

L.N.L.	Loyal North Lancs.
R.E.	Royal Engineers.
H.A.C.	Hon. Artillery Company.
R.F.A.	Royal Field Artillery.
M.G.C.	Machine Gun Corp.
K.R.R.	Kings Royal Rifles.
K.O.R.L.	King s Own Royal Lancs.
R.D.C.	Royal Defense Corps.
A.S.C.	Army Service Corps.
R.A.F.	Royal Air Force.

War Memorial . In Memory of the men from the Dick, Kerr Works, Preston who gave their lives in the Great War 1914–1919.

WARTON'S HERITAGE DEPARTMENT

The Heritage Department at Warton, previously known as the BAE Systems North West Heritage Group, was established in June 1995 by a group of volunteers following an appeal by the Company to help support its heritage policy. The Department covers the heritage of the Company in Lancashire, namely the sites at Warton, Samlesbury and formerly Preston.

The Department now has over 40 active volunteers whose range of activities include the establishment and management of document, photographic and film archives. The material held by the Department is used to promote the proud heritage of the Company at various displays and exhibitions as well as answering a constant stream of enquiries from within and outside the Company and producing heritage publications such as this one on the history of Dick, Kerr & Co. Ltd.

The Department are always keen to hear from past and present employees who may have material or personal stories that would be of interest.

Finally, do you have a story to tell that you feel would be a good subject for a future heritage publication? If so, please contact us and we will be pleased to discuss your ideas further.

BAE Systems Heritage Department member and writer John Shorrock has devoted several years to researching and writing '***Dick, Kerr, & Co. Limited***' at no cost to the project and all profits from sales of this and other Heritage Department publications are donated to BAE Systems Heritage Department - nominated local charities.

Ian Lawrenson
Heritage Manager - Warton
BAE Systems Heritage Department (W298)
Warton Aerodrome
Nr Preston
Lancashire
PR4 1AX
Tel: 07793 420193
Email: ian.lawrenson@baesystems.com

Heritage Department publications:

- The English Electric P.1 Supersonic Fighter Concept
  (Keith Emslie)
- The Wind Tunnel Department at Warton Aerodrome
  (Keith Emslie)
- Test Flying in Lancashire from Samlesbury and Warton Aerodromes:
  Volume 1: WW1 to the 1960s
  (James H. Longworth)
- Test Flying in Lancashire from Samlesbury and Warton Aerodromes:
  Volume 2: From the 1960s into the 1980s
  (James H. Longworth)
- Test Flying in Lancashire from Samlesbury and Warton Aerodromes:
  Volume 3: From the 1980s into the New Millennium
  (James H. Longworth)

Burnley Corporation Tram Fleet Number 41, on St. James Street, Burnley.
Manufactured by the Electric Railway and Tramway Carriage Works, Preston in 1903 Seating 44 passengers.